Alternative Fuels Guidebook

Properties, Storage, Dispensing, and Vehicle Facility Modifications

Richard L. Bechtold, P.E.

Society of Automotive Engineers, Inc.
Warrendale, Pa.

Library of Congress Cataloging-in-Publication Data

Bechtold, Richard L., 1952-
    Alternative fuels guidebook : properties, storage, dispensing, and
vehicle facility modifications / Richard L. Bechtold.
      p. cm.
    Includes bibliographical references and index.
    ISBN 0-7680-0052-1 (hardcover)
    1. Internal combustion engines, Spark ignition--Alternate fuels.
I. Title.
TP343.B37 1997
662'.6--dc21   97-27727
                                              CIP

Copyright © 1997 Society of Automotive Engineers, Inc.
               400 Commonwealth Drive
               Warrendale, PA 15096-0001
               U.S.A.
               Phone: (412) 776-4841; Fax: (412) 776-5760
               http://www.sae.org

ISBN 0-7680-0052-1

SAE Order No. R-180

Preface

I have been fortunate to be involved with alternative fuels for vehicles over the past 20 years. During that time, alternative fuels have evolved from experiments conducted in research laboratories to use by the public. I have personally been involved with vehicle modification to use alternative fuels, and more recently, design and installation of alternative fuel refueling facilities and modification of existing garages for safe storage and maintenance of alternative fuel vehicles. This experience includes underground and above-ground methanol refueling systems, compressed natural gas refueling systems, the Maryland Mass Transit Administration liquefied natural gas transit bus refueling system, and the Greater Richmond Transit Company electric bus recharging facility. The information herein includes the fundamentals that I have found to be essential to understanding the physical and chemical properties of alternative fuels and how they impact refueling system design and modifications of existing garages for safety purposes. As such, it is a combination of reference and general guide for engineers and fleet managers whose job is to implement alternative fuel vehicles.

R.L. Bechtold
April 1997

Table of Contents

Introduction

The term "alternative fuel" has been used to describe any fuel suggested for use in transportation vehicles other than gasoline or diesel fuel. In many ways, the current situation regarding fuels for transportation vehicles resembles the time in the early 1900s when vehicle buyers could choose among internal-combustion, steam, or electric vehicles. During this period, there were great debates about which fuels were best—even Henry Ford envisioned many of today's concerns about fuel availability and the environment by investigating the use of ethanol as a renewable, home-grown fuel whose production would benefit agriculture. The wide availability of inexpensive gasoline as a by-product of kerosene refining (for lighting purposes) was surely a large factor in the subsequent success of the internal-combustion engine in transportation vehicles. Both steam and electric vehicles had characteristics much desired by consumers at the time such as low noise and good driveability. However, drawbacks such as lengthy start-up time and complexity of operation (steam vehicles) and short driving range (electrics) conspired to limit their appeal. With the advent of the electric starter, internal-combustion-engine vehicles achieved a combination of economy, range, and ease of use that steam and electric vehicles could not match. From the demise of steam and electric vehicles, internal-combustion vehicles using gasoline and diesel fuel have enjoyed virtually complete dominance of the market. Ongoing development over time has resulted in very durable and reliable vehicles that are safe and less damaging to the environment than ever before.

The original impetus for development of alternative fuels to gasoline and diesel fuel was the realization that the oil-producing nations that held the majority of the world's reserves had the power[1] to dictate the price and availability of what

---

[1] The extent and completeness of this power is a topic of much debate, but there is no disagreement that the potential exists for much greater use of monopolistic powers if coordinated efforts by oil-producing nations are applied.

had become a truly international commodity. The U.S., several European countries, and Japan have led the development of alternative fuels because they have become dependent on oil imports to satisfy their transportation vehicle fuel needs. Today, the U.S. highway transportation sector is essentially totally dependent on petroleum fuels, making transportation very vulnerable to oil shortages and sudden price increases. The U.S. now uses more petroleum fuels in light-duty vehicles than *all* of its domestic oil production. Growth in light-duty vehicle fuel consumption is projected to be 0.8% through the year 2015, and for heavy-duty vehicles the fuel consumption growth rate is projected to be 1.3%. At the same time, growth in net oil imports is projected to be 1.9%, which predicts an ever-widening gap between petroleum consumption and domestic production.[2] Other petroleum-using sectors of the U.S. such as industry and utilities have made provisions to switch to fuels other than oil, while the transportation sector has not. It is this dependence on petroleum fuels that is prodding the use of alternative fuels in transportation.

Through experimentation with alternative fuels, it soon became clear that alternative fuels had inherent environmental advantages as well. Each alternative fuel has some characteristic that gives it an environmental advantage over petroleum fuels. Most are less damaging to the environment if spilled, and, in general, the emissions from alternative fuels are less reactive. This results in reduced amounts of ozone being produced with the benefit of improved air quality. In the 1980s there was less concern about energy security in the U.S., but the environmental advantages of alternative fuels kept interest high. In the 1990s the pendulum has swung back to the energy security value of alternative fuels. U.S. dependence on foreign oil, particularly for transportation, has assumed a steady increase approaching levels in the past when oil shortages or price shocks have occurred. Another reason interest in alternative fuels has again centered on energy security is because emission control technology combined with cleaner petroleum fuels such as reformulated gasoline and "clean diesel" has resulted in emission levels low enough to significantly depreciate the emissions benefits of alternative fuels.

---

[2] U.S. Department of Energy, Energy Information Administration, A1996 Annual Energy Outlook - 1996 - With Projections to 2015,@ DOE/EIA-0383(96), January 1996, National Energy Information Center, EI-231, Energy Information Administration, Forrestal Building, Room 1F-048, Washington, D.C. 20585.

The initial work on alternative fuels focused on which one was best from the viewpoint of technical feasibility, production capability, and cost. That question was never answered with certainty and, in the interim, development of alternative fuel vehicle technology has proceeded in parallel. Technical feasibility is no longer questioned, and the focus now has shifted more toward which alternative fuels can be produced at a competitive cost. Cost is calculated in terms not only of fuel price, but vehicle price and operating characteristics, and the expense of developing a national fuel distribution infrastructure. In addition, new issues such as public awareness and training of vehicle maintenance personnel have arisen as the use of alternative fuel vehicles spreads. Professions only peripherally aware of vehicle technology, such as professional engineers that must design vehicle storage and maintenance facilities, will need to become familiar with the physical characteristics and safe handling practices of alternative fuels. Building code and standards-setting organizations are slowly gaining the necessary information to address alternative fuels, though the process for change of codes and standards is a thorough one that takes many years to complete.

Alternative fuel vehicles will likely become more prevalent throughout the U.S. as a result of the passage and implementation of the Energy Policy Act of 1992, known as EPACT. EPACT requires the Federal government, state governments, and companies producing alternative fuels (fuel providers) to purchase alternative fuel vehicles as part of their new vehicle acquisitions. The Federal government has to date acquired approximately 15,000 alternative fuel vehicles, and the regulations for state and fuel providers to begin to acquire alternative fuel vehicles went into effect on March 16, 1996.[3] EPACT also includes provisions for requiring private and local fleets to purchase alternative fuel vehicles if it is determined that the petroleum displacement caused by Federal, state, and fuel provider alternative fuel vehicles is insufficient to meet the petroleum displacement goals of EPACT (if enacted, this mandate would take effect in 2002). The U.S. Energy Information Administration estimates that within ten years, annual sales of alternative fuel vehicles could exceed one million per year because of state mandates in addition to EPACT and from market-driven sales of alternative

---

[3] *Federal Register*, Vol. 61, No. 51, Thursday, March 14, 1996, p. 10622.

fuel vehicles.[4] These alternative fuel vehicles will create substantial demand for new fuel storage and dispensing facilities, and for modifications of existing facilities.

The objective of this book is to inform engineers and other interested parties about alternative fuels. It is directed at the professionals whose responsibilities require a working knowledge of alternative fuels, and who need a ready reference to inform and guide them in making decisions in their work. It concentrates on alternative fuels, their properties, characteristics, materials compatibility, and safe handling practices. It does not attempt to include the vehicle technology for using alternative fuels, nor their efficiency and emissions characteristics. These are changing very rapidly and are therefore not conducive for inclusion in a text meant to have more lasting content.

The alternative fuels included in this book are those which are considered the most likely candidates for use in internal-combustion engines and future energy conversion devices such as fuel cells. The alcohols (methanol and ethanol), natural gas (compressed and liquefied), LP gas, vegetable oils, and hydrogen are all covered in their entirety. Electricity is included only in terms of facility modifications for recharging and storing electric vehicles since the means for distributing electricity is not affected by its use in electric vehicles. Dimethyl ether (DME) is a promising alternative fuel for diesel engines made from natural gas, with physical properties similar to LP gas. Very little work has been done to define the production processes, typical composition data of DME from such plants, and storage and dispensing requirements for DME (though they are likely to be very similar to those for LP gas). At present, it is too early to provide guidance about how DME should be stored and dispensed.

---

[4] U.S. Department of Energy, Energy Information Administration, A1996 Annual Energy Outlook - 1996 - With Projections to 2015,@ DOE/EIA-0383(96), January 1996, National Energy Information Center, EI-231, Energy Information Administration, Forrestal Building, Room 1F-048, Washington, D.C. 20585.

Chapter One

Alternative Fuels and Their Origins

This chapter explains what alternative fuels are and why they are being considered for use in transportation vehicles. The production processes to make each are presented along with examples of recent production volumes. The typical impacts on vehicle performance (power, driveability, cold-start capability, etc.) are presented as well as typical vehicle emissions characteristics.

The Alcohols

Methanol and ethanol are the alcohols considered to be potential transportation alternative fuels. None of the alcohols higher than methanol and ethanol have been seriously considered as alternative fuels for use unmixed with other fuels in engines. Tertiary butyl alcohol (TBA) has been used as a gasoline extender and co-solvent when mixing methanol with gasoline, but not as a fuel by itself. Recently, dimethyl ether (DME, made using methanol) has been proposed for use as a diesel engine alternative fuel because of its very favorable emissions characteristics relative to using diesel fuel.

Methanol and ethanol make good candidates for alternative fuels in that they are liquids and have several physical and combustion properties similar to gasoline and diesel fuel. These properties are similar enough so that the same basic engine and fuel system technologies can be used for methanol and ethanol as for gasoline and diesel fuel. Both methanol and ethanol have much higher octane ratings than typical gasoline—which allows alcohol engines to have much higher compression ratios, increasing thermal efficiency. However, a significant drawback to methanol and ethanol relative to gasoline is that they have lower energy density, i.e., fewer Btus per gallon. It takes nearly two gallons of methanol and one and one-half gallons of ethanol to equal one gallon of gasoline. While this is not a problem for vehicles designed with larger fuel tanks, current gasoline vehicles that are configured for alcohol fuels do not have the latitude to increase fuel tank size significantly because of the intensive efforts to utilize all available space within a vehicle for competitive purposes.

Methanol and ethanol have inherent advantages relative to conventional gasoline and diesel fuel in that their emissions are less reactive in the atmosphere, producing smaller amounts of ozone, the harmful component of smog. Compared to specially formulated petroleum fuels such as California Phase 2 gasoline, the advantage is smaller. The mass of emissions using methanol or ethanol is not significantly different than from petroleum fuels. Methanol and ethanol have the disadvantage in that they produce formaldehyde and acetaldehyde as combustion by-products in larger quantity than the toxic compounds from the petroleum fuels that they replace.

Methanol and ethanol were long considered good spark-ignition engine alternative fuels. Clever implementation of ignition aids and use of fuel additives in the 1970s and 1980s proved that it was possible to use methanol and ethanol as diesel engine alternative fuels. A significant advantage of alcohol fuels is that when they are combusted in diesel engines, they do not produce any soot or particulates and they can be tuned to also produce very low levels of oxides of nitrogen. The U.S. Environmental Protection Agency (EPA) has determined that both particulates and oxides of nitrogen from heavy-duty diesel engines should be reduced much further in the interests of the health of the populace [1.1].

Methanol

Consideration of methanol as motor fuel did not emerge until it became a common industrial chemical. It was used as an automotive fuel during the 1930s to replace or supplement gasoline supplies, in high-performance engines in Grand Prix racing vehicles in the 1930s, and for about the last two decades as the only fuel allowed in competing vehicles at the Indianapolis 500 [1.2, 1.3]. For general use, serious research attention started in the late 1960s based on emissions advantages and was greatly expanded when energy security problems developed in the 1970s. Air quality has become the near-term catalyst for alternative fuel vehicle expansion, but in the long-term the United States and the rest of the world will have to rely on alternative sources of energy as petroleum reserves diminish [1.4]. With the U.S. government-mandated phase-out of lead as a gasoline octane additive, low concentrations of methanol were found to be a good nonmetallic substitute. Methanol requires incorporation of higher-order (C_3 - C_8) alcohols to obviate phase separation deficiencies, and a 50/50 mixture with TBA was found to be more advantageous than other organics [1.5]. Over time the use of methyl tertiary butyl ether (MTBE) replaced blends of methanol and TBA. (MTBE is made by reacting methanol with iso-butylene.) MTBE is an octane blending agent with more favorable characteristics than methanol to produce blends with gasoline for use in existing gasoline vehicle models [1.6].

Methanol's major advantages in vehicular use are that it is a convenient, familiar liquid fuel that can readily be produced using well-proven technology. It is a fuel for which vehicle manufacturers can, with relative ease, design a vehicle that will outperform an equivalent gasoline vehicle and obtain an advantage in some combination of emission reduction and efficiency improvement [1.7].

Major disadvantages of methanol are: initial higher cost than that of gasoline; impact of reduced energy density on driving range or larger fuel tank; it burns with a flame that is not visible in direct sunlight; and need for education of users and handlers on toxicity and safety [1.8].

Production

Methanol is a colorless liquid that is a common chemical used in industry as a solvent and directly in manufacturing processes. In 1995, methanol production in the U.S. totaled 6.5 billion liters (1.7 billion gallons), which made it the 21st ranked chemical in terms of use [1.9]. One of the largest single uses of methanol is to make MTBE, the preferred oxygenate for addition to gasoline for wintertime oxygenate programs[1] and to make reformulated gasoline. In 1995, 10.0 billion liters (2.64 billion gallons) of MTBE was produced in the U.S., making it the 12th most-used chemical [1.9].

---

[1] The U.S. Environmental Protection Agency requires that MTBE or other oxygenates be added to gasoline in certain areas of the U.S. during the winter to reduce emissions of carbon monoxide.

Methanol was once referred to as "wood alcohol" because it originally was made from the destructive distillation of wood. The technology for large-scale production of methanol was developed by Badische Anilin und Soda Fabrik (BASF) in Germany in 1924 [1.8]. The currently preferred process for producing methanol is steam reformation of natural gas. In this process, any sulfur present in natural gas is first removed. Next, the natural gas is reacted with steam in the presence of a catalyst under high heat and pressure to form carbon monoxide and hydrogen. These elements are then put through the methanol production catalyst to make methanol. There are many variations of the basic steam reforming process, all aimed at increasing the overall thermal efficiency. Steam reformation of natural gas has a thermal efficiency of about 56-62%, while advanced processes can have thermal efficiencies as high as 68%. Figure 1-1 shows a typical methanol plant in use today.

Larger methanol production plants are more efficient than smaller ones. The size of a large (called world-scale) methanol plant is in the range of 2000-2500 metric tons per day. If methanol were to become a widely used alternative fuel, many more methanol production plants would be required. Plants as large as 10,000 metric tons per day have been postulated to serve the demand created by transportation vehicles.

Methanol can also be produced from coal and municipal waste. However, only one plant in the U.S. now produces methanol from coal.

In the U.S., the primary methanol production location is in the Gulf Coast area. Methanol is also produced in Canada, South America, Europe, and the Middle East. Methanol production and price is not controlled by any single country or consortium of countries. Any country with remote natural gas reserves is a candidate for methanol production since production of methanol usually represents the most cost-effective means of developing those reserves.

Methanol is distributed throughout the U.S. as an industrial chemical. Most suppliers can be found in the phone book under chemical or industrial supplies. The price of methanol in small quantities is much higher than if delivered in large bulk quantities (such as how petroleum fuels are delivered).

Fig. 1-1 Methanol Production Plant.
(Source: Caribbean Methanol Company)

Vehicle Emissions Characteristics

Methanol burns without a visible flame, which is a safety concern, but which also demonstrates that methanol does not produce soot or smoke when combusted. This fact makes methanol a very attractive diesel engine fuel because, unlike diesel fuel, no fuel particulates are formed. Recent health effects studies suggest that particulate matter is a health hazard regardless of whether known carcinogenic hydrocarbons are absorbed onto the particulate or not [1.10]. Though methanol is poisonous, diesel fuel contains several hydrocarbons that are known or suspected carcinogens. Methanol exposure studies have shown that methanol does not cause harm in the quantities that would accumulate in the body from exposure from refueling vapors or from unburned methanol in vehicle exhaust. In addition, because of methanol's high latent heat of vaporization, peak combustion temperatures can be reduced with correspondingly low emissions of oxides of nitrogen (NO_x). This, combined with the fact that methanol is less reactive in the atmosphere than hydrocarbons from diesel fuel make the emissions benefits of using methanol in diesel engines very positive. The NO_x emissions of the Detroit Diesel Corporation 6V-92TA methanol engine[2] certified in 1993 of 1.7 gm/Bhp·hr have yet to be bettered by conventional diesel engines or diesel engines converted to natural gas operation.

The physical and chemical properties of methanol can be used advantageously in engines to produce low emissions. Because it is less photochemically reactive than gasoline, its evaporative emissions contribute less to smog formation; and because it contains oxygen, it facilitates leaner combustion resulting in lower CO emissions. The higher latent heat of vaporization results in lower combustion temperatures, reducing generation of NO_x [1.11]. Evaporative emissions of methanol during transport, storage, dispensing, and use fall about midway between gasoline and diesel fuel, but increase with use of gasoline/methanol blends. Even though nearly twice as much methanol by volume is required to achieve the same operating range as gasoline, evaporative losses from M100[3] distribution could be about two-thirds those of gasoline [1.12]. Unburned fuel (hydrocarbons) is less reactive because it is primarily methanol. Methanol's low specific reactivity means that unburned methanol and evaporated methanol emissions have less smog/ozone-forming potential than an equal weight of organic emissions from gasoline-fueled vehicles and infrastructure [1.7]. However, this advantage is offset somewhat by the formaldehyde that methanol produces when combusted. Emissions of M85[4] in spark-ignition light-duty vehicles have been shown to have significant reductions in specific ozone-forming potential (about 50% grams ozone

---

[2] This engine is no longer in production.

[3] "M100" is used to represent pure, 100%, or "neat" methanol.

[4] "M85" is used to represent a fuel blend of 85% methanol and 15% gasoline. The gasoline portion is added for cold-start and flame luminosity reasons.

per gram emissions) compared to conventional gasoline (40-45% reduction compared to California Phase 2 reformulated gasoline) [1.13]. Some tests of early model methanol flexible fuel vehicles (see next section) showed increased emissions per mile relative to using gasoline, which offset the reactivity advantage of methanol. Tests of more recent model flexible fuel vehicles have shown that they can achieve very similar emissions per mile as when using gasoline [1.14].

Methanol contains no sulfur, so it does not contribute to atmospheric sulfur dioxide (SO_2). Since SO_x and NO_x emissions lead to acidic deposition, use of methanol would make a minor contribution to reducing acid rain [1.15]. The CO_2 emissions of methanol vehicles are theoretically about 94% those of similar petroleum-fueled vehicles, assuming they have the same fuel efficiency. However, producing methanol (steam reformation of natural gas) releases almost half the greenhouse gases that producing gasoline does. When the entire fuel cycle from resource through combustion is included, methanol has very similar greenhouse gas emissions relative to gasoline [1.16, 1.17].

Vehicle Performance Impacts

Methanol spark-ignition engines have the capability to be 15-20% more efficient than their gasoline counterparts. This is achieved through lean-burn technology, made possible by methanol's wide flammability limits. Besides having superior thermal efficiency, lean-burn engines have lower exhaust emissions with simpler oxidation-catalyst technology. HC and CO emissions have been demonstrated to be much lower from lean-burn vehicles, with NO_x emissions about the same as those from current gasoline vehicles [1.3].

Ford built 630 dedicated methanol Escorts from 1981 to 1983. The engine in these vehicles had 20% higher torque than the original gasoline versions. Acceleration from 0 to 60 mph was 15.8 seconds on M85 compared to 16.7 seconds on gasoline. Subsequent Crown Victorias had an acceleration time of 12.2 seconds on M85 compared to 12.6 seconds on gasoline [1.12].

Flexible fuel vehicles (FFVs) are able to operate on gasoline, M85, or any mixture of the two fuels. This is accomplished through the use of a fuel composition sensor in the fuel line from the vehicle fuel tank to the engine. The engine control system automatically adjusts the air-fuel ratio and the spark timing for the blend of methanol and gasoline traveling to the engine. This system allows the use of only one fuel tank and the driver need not be concerned about mixing two fuels in any ratio. Thus, the driver can use methanol when available, or gasoline when methanol is not available or when the vehicle travels outside the range of methanol refueling facilities.

Since the FFV offers the consumer freedom from dependency on a novel fuel and also the option to choose that fuel if it is available at the right price, the majority of light-duty methanol vehicles sold for the next decade or more probably will be FFVs. Since M85 differs from gasoline in several significant ways, the FFV will require several special features that will increase its production cost by $100-$400 over the price of a comparable conventional vehicle.

The heart of the FFV design is a special sensor located in the fuel line between the fuel tank and the engine. This sensor instantaneously (under 50 milliseconds) determines the concentration of methanol (and/or ethanol) in the fuel. This information is fed to the engine computer which calculates appropriate spark advance and the pulse width of the fuel injector signal for correct fuel volume. Figure 1-2 illustrates the differences in Ford's 1996 Ford Taurus FFV compared to gasoline versions of their Taurus. Ford has built this FFV in models optimized for M85 and for E85[5].

FFVs do leave room for improvement as General Motors demonstrated by optimizing a 3.1L Lumina to dedicated M85 operation. The engine compression ratio was increased from 8.9 (used in the FFV version) to 11.0. This was obtained by use of a smaller piston bowl, reduced crevice volume, etc. A close-coupled 80-cu.-in. palladium catalyst was added to the 170-cu.-in. palladium/ rhodium one. Performance results as compared to the FFV predecessor are shown in Table 1-1 [1.17]. Not only was performance increased, but fuel economy was increased while emissions were decreased. This vehicle was even then very close to meeting the California ULEV emissions standards.

---

[5] "E85" is used to represent a blend of 85% ethanol and 15% gasoline. The gasoline is added primarily for cold-start purposes, and contributes to greater flame luminosity.

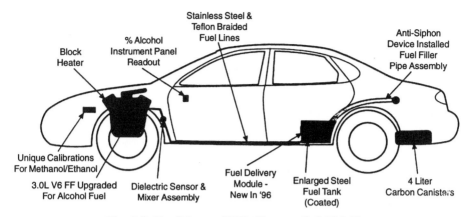

Fig. 1-2 Ford Taurus FFV. (Source: Ref. [1.14])

Table 1-1. Performance Results

Characteristic	FFV	Dedicated
Acceleration, 0-60 mph (sec)	10.4	9.4
Quarter mile (sec)	17.7	17.2
Fuel economy (mpg[a])	36.6	37.9
Emissions (g/mi)		
OMHCE	0.15	0.06
CO	2.06	0.80
NO_x	0.16	0.22
Formaldehyde (mg/mi)	14.4	5.7

Dedicated Emissions (50,000 mile) in comparison to:

	TLEV	LEV	ULEV
NMOG	33%	55%	103%
CO	24%	24%	50%
NO_x	55%	110%	110%
HCOH	38%	38%	71%

a Gasoline Equivalent

Ethanol

Ethanol has long been considered a good spark-ignition engine fuel, and engines were run on ethanol very early in engine development. Henry Ford was an early proponent of using ethanol as a fuel because of its good combustion properties and because of its potential "self-sufficiency," i.e., it can be produced by the agricultural sector which would satisfy their needs and sell the excess to others. Brazil, in fact, has implemented this idea and is the only country around the world to have done so to date.

Production

Ethanol produced for use as a fuel (hereafter called fuel ethanol) in the U.S. is produced almost exclusively using fermentation technology. The preferred feedstock is corn, though other grains and crops such as potatoes and beets can be used. Agricultural wastes such as cheese whey are also considered good feedstocks for ethanol production. Almost any source of starch or sugar is a potential feedstock for ethanol production. (Starches are saccharified to sugars, which are then fermented.) In Brazil, sugar cane is the preferred feedstock for ethanol production because of a favorable growing climate. In France, ethanol is produced from grapes that are of insufficient quality for wine production.

What governs the choice of feedstock is cost and the capability for large-scale production. Included in the cost is the amount of petroleum used to produce the crop and then prepare it for fermentation. The petroleum used to produce ethanol reduces the petroleum displacement value of ethanol as an alternative fuel. Petroleum use for ethanol production varies widely depending on the area of the country, the crop grown, and the agricultural practices used. Should ethanol become available as an alternative fuel in large volumes, it would be expected that it would be used instead of petroleum to fuel the agricultural equipment used in crop production thus increasing its petroleum displacement value.

There are three primary ways that ethanol can be used as a transportation fuel: 1) as a blend with gasoline, typically 10% and commonly known as "gasohol"; 2) as a component of reformulated gasoline both directly and/or transformed into a compound such as ethyl tertiary butyl ether (ETBE); or 3) used directly as a fuel, with 15% or more of gasoline known as "E85." Ethanol can also be used directly in diesel engines specially configured for alcohol fuels, such as the Detroit Diesel Corporation 6V-92TA model.[6] Using ethanol to make gasohol or in reformulated gasoline, or transformed into ETBE for use in reformulated gaso-

---

[6] This model of engine is no longer in production.

line, does not require specially configured vehicles; almost all current vehicles will tolerate these fuels without problem and with likely advantageous emissions benefits. For example, ethanol and ETBE blends in gasoline are approved for use by the Environmental Protection Agency (EPA) for mandated winter-time oxygenated fuel programs whose objective is to lower vehicle carbon monoxide emissions.

The U.S. currently has a production capacity for fuel ethanol of 1.1 billion gallons per year. Ninety percent of this capacity is from 16 plants of 10 million gallons per year and larger. Archer Daniels Midland accounts for 700 million gallons per year of this capacity, and 12 other producers are represented by the remaining production capacity. About 690 million gallons per year of new production capacity is planned, and 200 million gallons per year of production is currently shut-in, though it is estimated that only 121 million of that capacity is feasible to be reactivated [1.18, 1.19]. Almost all ethanol production plants are located in the Midwest where the largest amount of corn is grown.

Fig. 1-3 Ethanol Production Plant. (Source: Reeve Agri Energy, Garden City, Kansas)

Fuel ethanol production has increased rapidly during the 1980s, spurred by a $0.54 per gallon Federal tax subsidy. In 1978, only 10 million gallons of fuel ethanol was sold. In 1991, the U.S. produced 875 million gallons of fuel ethanol that was all used in the U.S. except for about 50 million gallons that was exported to Brazil. The U.S. imports about 25 million gallons per year of ethanol from the Caribbean countries. The large increase in fuel ethanol capacity is due almost entirely to the construction of world-scale fuel ethanol production facilities. Smaller fuel ethanol production facilities do not have the economies of scale to produce fuel ethanol at prices competitive with large facilities. Expansion of fuel ethanol production capacity depends primarily on the value of ethanol relative to petroleum fuels, which in turn depends on tax credits for fuel ethanol production.

Vehicle Emissions Characteristics

Ethanol by itself has a very low vapor pressure, but when blended in small amounts with gasoline, it causes the resulting blend to have a disproportionate increase in vapor pressure. For this reason, there is interest in using fuels such as ETBE as reformulated gasoline components because ETBE has a small blending vapor pressure (28 kPa; 4 psi) which will reduce the vapor pressure of the resulting blend when added to gasoline. (Economics have been holding back use of ETBE relative to MTBE.) The primary emission advantage of using ethanol blends is that CO emissions are reduced through the "blend-leaning" effect that is caused by the oxygen content of ethanol. The oxygen in the fuel contributes to combustion much the same as adding additional air. Because this additional oxygen is being added through the fuel, the engine fuel and emission systems are "fooled" into operating leaner than designed, with the result being lower CO emissions and typically slightly higher NO_x emissions. The blend-leaning effect is most pronounced in older vehicles that do not have feedback control systems; however, even the newest technology vehicles typically show some reduction in CO emissions.

The emission characteristics of E85 vehicles are not as well documented as for M85 vehicles; however, Ford tested E85 in their 1996 model Taurus flexible fuel vehicle and found essentially no difference in tailpipe emissions compared to using the standard emissions testing gasoline (Indolene). In this test, the engine-out emissions of HC and NO_x were lower than for gasoline, but ethanol's lower exhaust gas temperatures were believed to decrease catalyst efficiency slightly so that the tailpipe emissions were the same. E85 produces acetaldehyde instead of formaldehyde when methanol or M85 is combusted, and Ford found that the level of acetaldehyde was the same as for M85 [1.14]. An advantage of acetaldehyde over formaldehyde is that it is less reactive in the atmosphere which contributes less to ground-level ozone formation. The National Renewable Energy Laboratory has demonstrated that a 1993 Ford Taurus FFV could be modified so that the stringent California ULEV emissions regulations (the most stringent regulations for light-duty vehicles) could be met using E85 as fuel [1.20]. The engine, fuel system, and emissions system changes made to achieve ULEV emissions levels are not specific to the Taurus FFV and could be applied to most any other E85 vehicle. The specific reactivity of ethanol exhaust emissions (grams ozone per gram emissions) has been measured to be significantly lower than conventional gasoline (about 30%) and lower than from specially formulated gasoline (California Phase 2) [1.21].

The low sulfur content of E85 should be a benefit in reducing catalyst deterioration compared to vehicles using gasoline. Insufficient data have been gathered to date to determine whether this effect is significant.

The vehicle technology to use E85 is virtually the same as that to use M85, allowing auto manufacturers to quickly build E85 vehicles since they have already engineered M85 vehicles. General Motors introduced a Chevrolet Lumina E85 FFV in the 1992 model year and sold 386 in 1992 and 1993. Ford introduced its E85 FFV Taurus for the 1996 model year and produced between 2500 and 3000 after many years of producing M85 FFV Tauruses. This rapid introduction of E85 vehicles could not be possible without the pioneering work that was done on M85 vehicles first.

Vehicle Performance Impacts

When ethanol is used as a blending component in gasoline there are typically few discernable differences in vehicle driveability or performance to the driver of current technology vehicles. (Some older vehicles have experienced hot-start problems because of increased volatility, and vehicles using ethanol blends after many years of gasoline use may experience fuel filter plugging because the ethanol acts as a solvent for gasoline deposits.) FFVs built to use E85 should experience very similar driveability as when using gasoline, though performance should be improved by about 5% because of the intake charge cooling effects (high latent heat of vaporization) and high octane number of ethanol. Fuel efficiency in energy terms should be the same or slightly better when using E85 in an FFV, since the combustion properties of ethanol should favor a more aggressive spark timing program without changes in emissions. An FFV optimized for use with E85 through the use of increased compression ratio should have even higher performance, greater fuel efficiency, or a combination of both, depending on calibration of the engine control system.

The maintenance requirements of E85 vehicles should be essentially the same as for M85 vehicles, and very similar to their conventional petroleum fuel vehicle counterparts. The oil specified for E85 vehicles has a special additive package and is currently expensive (approximately $3.00 per quart) because of low volume production. In high volume production E85 engine oil should be no more expensive than gasoline engine oil.

There is no reason to believe that E85 vehicles should not last as long as gasoline vehicles. Long-term tests of M85 vehicles have been very successful and have shown similar engine wear to the same engines using gasoline. This should hold true for E85 vehicles as well.

Natural Gas

Natural gas is one of the world's most abundant fossil fuels and currently supplies over 25% of the energy demand in the U.S. as shown in Figure 1-4 [1.22]. The primary uses of natural gas in the U.S. are for space heating, electricity generation, and industrial processes. Natural gas is a very good spark-ignition internal-combustion engine fuel and it was used as a fuel in the very early days of engine development. However, relative to liquid petroleum fuels, the ability to store sufficient amounts of natural gas for onboard vehicles has presented a significant barrier to its broad use as a transportation fuel. Significant advances have been made in high-pressure cylinders that can store natural gas at high pressures (up to 3600 psi) that are made of lightweight materials including aluminum and carbon fiber. Compressed natural gas (CNG) is the preferred method of natural gas storage on vehicles. Liquefied natural gas (LNG) is gaining favor for use in heavy-duty vehicles where use of CNG would still entail space and load-carrying capacity penalties. Storing natural gas as LNG enables heavy-duty vehicles to have the same operating range as when using liquid petroleum fuels. Typically, storing natural gas as LNG instead of CNG results in a fuel storage

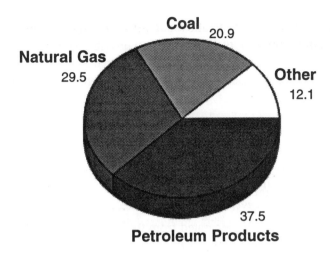

Fig. 1-4 U.S. Energy Consumption, %, by Source.
(Source: Energy Information Administration, "Annual Energy Outlook, 1996")

system that is less than half the weight and volume of a CNG system. Regardless of the method of storage, the cost and emissions advantages of natural gas make it a very popular alternative fuel.

Production

Natural gas is present in the earth and is often produced in association with production of crude oil. However, wells are also drilled for the express purpose of producing natural gas.

The main constituent of natural gas is methane, the lightest and simplest hydrocarbon, composed of one carbon and four hydrogen atoms. Ethane is typically the only other hydrocarbon found in significant amounts in natural gas, though often less than 10 volume percent. Natural gas may also include carbon dioxide, nitrogen, and very small amounts of hydrogen and helium.[7] The composition of natural gas is important because its heating value and physical properties may change which can affect combustion.

---

[7] Helium is produced primarily by separating it from natural gas. Natural gas as delivered by utilities may also include propane and air added to stretch the supply of natural gas during periods of high demand, typically during winter months.

The properties of natural gas are dominated by methane. Methane is widely acknowledged to be formed from four sources: 1) organic matter that is decomposed in the presence of heat; 2) organic matter that is converted through the actions of microorganisms; 3) oil and other heavy hydrocarbons that produce methane in the presence of heat; and 4) coal which releases methane over time [1.23]. There is a theory that methane is present in large quantities deep within the earth, from which it migrates upward via cracks and fissures. This theory, known as the abiogenic theory, is not proven but if found true would suggest that very large reserves of methane exist in the earth.

Very large reserves of natural gas are believed to lie at depths of 4600-9200 meters (15,000-30,000 feet), called "deep gas." Since methane remains stable up to its autoignition temperature of 550°C (1022°F), it is found at depths where oil is not found, presumably because oil will be transformed in part to methane at lower temperatures. Deep gas is expensive to drill for, but the quantities are estimated to be very large. Technology has been developed to enhance recovery of deep gas when it is found.

Very little processing needs to be done to natural gas to make it suitable for use as a fuel. Water vapor, sulfur, and heavy hydrocarbons are removed from natural gas before it is sent to its destination, usually via pipeline. Compared to liquid hydrocarbon and alternative fuels, natural gas contains much less energy per unit volume. For this reason, transport over long distances and across oceans is not practical except when liquefied. Liquefied natural gas is created when natural gas is cooled to −127°C (−260°F), using a variety of techniques. Even when liquefied, LNG contains only approximately 60% the energy of gasoline per unit volume, and while the technology of insulating containers has advanced significantly, loss of LNG due to heat transfer into the containers eventually occurs. Even so, LNG is the only practical way remote sources of natural gas can be transported long distances and across oceans.[8]

Vehicle Emissions Characteristics

Natural gas is composed primarily of methane which dominates its emissions characteristics. Methane mixes readily with air and has a high octane rating

---

[8] Another method of making economical use of remote natural gas reserves is to produce methanol from them. Methanol can easily be transported via ocean tanker without losses, unlike LNG.

which makes it a very good spark-ignition engine fuel. It has a high ignition temperature that makes it unsuited for use in compression-ignition engines, though it can be made to work in such engines. Methane barely participates in the atmospheric reactions that produce ozone, though it does contribute to global warming when released into the atmosphere. Because of its high hydrogen-to-carbon ratio (the highest at 4 to 1), the combustion of methane produces about 10% less carbon dioxide than combustion of the energy-equivalent amount of gasoline or diesel fuel. (Carbon dioxide is the primary contributor to global warming.) The emissions characteristics of both light- and heavy-duty vehicles are presented in the following subsections.

Light-Duty Vehicles

Light-duty CNG vehicles are capable of very low gaseous exhaust emissions. In the CNG vehicles developed by the auto manufacturers to date, individual port fuel injection with a three-way catalyst system has been used to simultaneously oxidize exhaust hydrocarbons and carbon monoxide while reducing oxides of nitrogen. Because natural gas readily mixes with air, emissions of carbon monoxide are typically low when using natural gas, assuming that the air-fuel ratio is kept on the lean side of stoichiometric. Low carbon monoxide emissions degrade the efficiency of oxides of nitrogen reduction in three-way catalyst systems and the auto manufacturers have resorted to biasing the air-fuel ratio slightly to the rich side to compensate. These changes, plus optimization of the catalyst precious metals for oxidation of methane, have resulted in vehicles with very low exhaust emissions. For example, Chrysler has been able to certify its CNG minivan to California's Ultra Low Emissions Vehicle standard, the most stringent in the U.S. for internal-combustion-engine vehicles.

In addition to very low exhaust emissions, CNG vehicles also have the advantage of no evaporative emissions or running loss emissions[9] caused by the fuel. As vehicle exhaust emissions are reduced to meet the most stringent standards, running loss emissions become more important.

---

[9] Running loss emissions are emissions from the fuel tank during vehicle operation due to heating of the fuel, losses from the evaporative canister, or losses due to permeation of fuel through elastomeric fuel lines.

Figure 1-5 illustrates the CNG fuel system used in the 1996 dedicated CNG Ford Crown Victoria.

Heavy-Duty Vehicles

The heavy-duty engine manufacturers have taken a slightly different approach to natural gas engine development compared to light-duty engine manufacturers. Heavy-duty natural gas engines to date have been spark-ignition adaptations of diesel engines. To improve engine efficiency to be closer to that of diesel engines, the heavy-duty natural gas engines use lean-burn combustion.

For emissions control, they use an oxidation catalyst to control methane and carbon monoxide emissions. Oxides of nitrogen are kept low through lean combustion and particulate emissions from natural gas are not a concern. Oxides of nitrogen emissions from heavy-duty diesel engines are an increasing concern with the U.S. EPA. The EPA has just recently proposed oxides of nitrogen emissions standards that will be difficult for diesel engines using diesel fuel to comply with, though natural gas heavy-duty engines will be able to fairly easily comply. If such emissions standards are put into place, the heavy-duty natural gas engine will likely be more competitive in terms of emissions and engine efficiency.

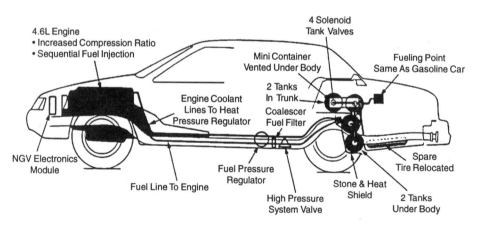

Fig. 1-5 Schematic of the CNG Fuel System in the 1996 Ford CNG Crown Victoria. (Source: Ref. [1.36])

Heavy-duty vehicles using natural gas favor storing natural gas as LNG instead of CNG. LNG vehicles should have the same exhaust emissions as CNG vehicles except that LNG vehicles might also occasionally vent methane from the fuel storage system. If LNG vehicles are used regularly, no venting of methane is necessary. However, if an LNG vehicle is left idle for a week or more, it will need to start venting to prevent excessive pressure in the fuel tanks.

Vehicle Performance Characteristics

Light-Duty Vehicles

Light-duty vehicle engines using natural gas can increase their power and efficiency by increasing the compression ratio. Compared to typical gasoline, natural gas has a high octane rating that will support higher compression ratios. However, there are two detriments to light-duty natural gas vehicle performance: the weight of the fuel system, and decreased engine specific power output. The weight of natural gas fuel systems will always be more than a liquid fuel system carrying the energy equivalent amount of gasoline. This weight naturally hurts vehicle acceleration and will degrade fuel economy proportionately. Since natural gas enters the engine entirely as a gas, while gasoline typically enters the engine as part liquid and part gas, for the same displacement, the gasoline engine will be able to ingest more air and fuel and produce higher power. Natural gas engines can overcome this inherent disadvantage by increasing compression ratio, cylinder displacement, or increasing volumetric efficiency. However, it is not always possible to obtain the same power output without creating a more expensive engine.

Natural gas light-duty vehicles should have the same good driveability characteristics that gasoline vehicles have obtained and customers expect.

Heavy-Duty Vehicles

Since heavy-duty natural gas engines are derivatives of heavy-duty diesel engines, it is possible for them to have the same or more power than the equivalent displacement diesel version. Limiting factors to power output include oxides of nitrogen emissions (increased power means richer operation and generation of more oxides of nitrogen for a given displacement engine) and exhaust valve life (spark-ignition engines experience higher exhaust valve temperatures than diesel engines). Rather than acceleration, heavy-duty engine performance is more a function of maximum horsepower and "torque rise." Both characteristics can be made equivalent or better for natural gas versions of heavy-duty diesel engines.

LP Gas

LP gas, or LPG, stands for liquefied petroleum gas. LP gas includes several light hydrocarbons whose main distinguishing characteristic is becoming a liquid when put under modest pressures (less than 300 psi). Propane and butane are the most common LP gases and, for vehicle use, LP gas is essentially all propane and is most often called propane.

Propane has been used as a vehicle fuel for at least the past 60 years in the U.S. It is by far the alternative fuel used in the largest volume in the U.S., estimated to be 97% of all alternative fuel use in 1992. The number of vehicles using propane as fuel has been estimated to be between 220,000 and 370,000 [1.24, 1.25]. Nearly all these propane vehicles are conversions of gasoline vehicles, since the auto manufacturers have sold only very few propane vehicles. Even so, the consumption of these propane vehicles accounted for just 0.1% of all the gasoline-equivalent on-road motor fuel use in 1992.

Propane is stored under pressure to keep it liquid. Unlike LNG, propane does not need anything other than a modest pressure to keep it liquefied. The tanks that store propane are pressure vessels, but since they can be made from low-carbon steel their price is modest compared to CNG and LNG tanks. The propane is vaporized in a device called the converter that lowers the pressure of the propane to vaporize it. The converter also uses engine coolant to warm the propane to ensure that it is completely vaporized when it passes on to the mixer. As its name suggests, the mixer mixes the propane and air in the desired ratio before it enters the engine. Other controls and compensation for temperature are included in the propane fuel system. Propane fuel systems can also incorporate feedback control to work with three-way catalyst emission control systems.

Production

About half of LP gas is produced in association with production of natural gas and half is produced in association with crude oil refining. It is undesirable to leave LP gases in natural gas as they come out of the ground because of their tendency to liquefy under modest pressures, such as those that exist in natural gas pipelines.

The major use of LP gas is for space heating in homes and for commercial purposes, though it is also an important feedstock for petrochemicals. Propane is a popular barbecue grill fuel and many forklift trucks operated inside warehouses use propane. The production of LP gas is fairly constant over the year, though consumption is highest during the winter months because of the large amounts used for space heating.

The total production of LP gas in the U.S. was about 20% of the volume of gasoline-equivalent[10] on-road motor fuel in 1992, or 15% on an energy basis. Of this amount, propane represents about 60%. If all propane were used as vehicle fuel, it could represent nearly 10% of the gasoline-equivalent on-road fuel used by vehicles. While this is likely not practical because of the large home heating market for propane, with the proper fuel systems, LP gases other than propane (the most prominant being butane) could also be used as vehicle fuel with appropriate engine modifications.

Vehicle Emissions Characteristics

Propane shares several emissions advantages with natural gas and has some additional ones of its own. Like natural gas, propane vehicles do not have any evaporative or running loss emissions associated with the fuel. Unburned hydrocarbons from propane are easier to oxidize in oxidation catalysts than methane, which results in low unburned hydrocarbon emissions. Unburned hydrocarbons from propane are also less reactive by about two-thirds compared to unburned hydrocarbons from gasoline [1.26]. Being a gas, propane mixes very well with the air before entering the engine, resulting in low carbon monoxide emissions,

---

[10] "Gasoline-equivalent" refers to equal energy in terms of gallons of gasoline. For instance, diesel fuel has about 10% more energy per gallon than gasoline, so it represents 1.1 gallons gasoline-equivalent.

assuming that the fuel system does not allow an overall rich mixture. When using a three-way catalyst, oxides of nitrogen emissions can also be reduced to very low levels if the air-fuel ratio is kept at the stoichiometric value.

One area where propane fuel systems could use improvement is in refueling emissions. A significant trapped volume of fuel can exist within the refueling connector when refueling is completed. This fuel is then released when the refueling connection is disconnected, releasing the trapped propane to the atmosphere. While all propane refueling connections are the same diameter and are interchangeable, the amount of trapped propane released varies due to detail differences among the refueling connectors. This situation should change starting in 1998 when the U.S. Environmental Protection Agency will require that the amount of propane released when refueling connections are broken be limited to less than 2 cc, which is approximately the same amount lost during refueling gasoline vehicles [1.27]. Vehicle manufacturers will have three years to phase in the use of new propane refueling connectors (starting in the 1998 model year for passenger cars and the 2001 model year for light-duty trucks). Refueling facilities must have the new refueling connections by January 1, 1998. Low-volume facilities (less than 10,000 gallons of gasoline equivalent per month) can obtain an extension until January 1, 2000.

Light-Duty Vehicles

Propane readily changes to a gas when its pressure is reduced, and it mixes readily with air before it enters the combustion chamber. This good mixing typically results in very low carbon monoxide emissions, assuming that the overall air-fuel ratio is kept on the lean side. Vehicles that have catalytic converters have also shown very low hydrocarbon emissions, presumably because unburned propane is more easily oxidized than the typical mix of unburned hydrocarbons produced from gasoline. Oxides of nitrogen emissions are typically unaffected compared to operation using gasoline. However, propane fuel systems that are calibrated to the lean side of stoichiometric may cause an increase in oxides of nitrogen, especially in those vehicles having three-way catalysts.

While the "jury" may still be out on the exhaust emissions of propane vehicles, it is certain that propane vehicles do not have evaporative or running loss emissions caused by the fuel. As exhaust emissions of vehicles become smaller and smaller, evaporative and running loss emissions become more important. This has positive implications for propane vehicles.

Heavy-Duty Vehicles

To date very few propane heavy-duty vehicles have been built and sold. Those that have been are typically at the small end of the heavy-duty vehicle scale that use large spark-ignition engines which were designed for gasoline (e.g., the Ford F600/F700 series propane trucks). These propane vehicles should have similar emissions compared to light-duty propane vehicles. Some heavy-duty engine manufacturers (notably DDC, Caterpillar, and Cummins) are actively research-ing and testing propane versions of their natural gas heavy-duty engines. It is probable that these propane engines will have similar emissions characteristics compared to the natural gas version of these engines.

Vehicle Performance Characteristics

The vast majority of propane fuel systems used on light-duty vehicles to date have been of the mechanical-control type that meter propane in proportion to the amount of air used by the engine (air-valve and venturi-type "mixers"). While these systems work well, their capabilities have been overshadowed by gasoline fuel injection systems and often lag behind gasoline systems in terms of accelera-tion, driveability, and cold-start performance. Chrysler Canada and one Euro-pean equipment manufacturer offer liquid propane injection systems that are direct analogs to gasoline port fuel injection systems. These systems should have inherent performance advantages compared to the vaporized propane fuel sys-tems.

Light-Duty Vehicles

For vaporized propane fuel systems, propane enters the engine as a gas instead of part liquid and part gas as gasoline does. By entering the engine fully vaporized, some air that could otherwise be used for combustion is displaced. Therefore, theoretically, propane vehicles should have lower power and slower acceleration than their gasoline counterparts, especially in bifuel[11] configuration. In practice,

---

[11] "Bifuel" vehicles are those with two fuel systems but can operate using only one at a time.

however, these differences do not always show up though power and acceleration are typically reduced. Acceleration in terms of 0-50 mph time can be up to 10% slower. The change in acceleration varies with some systems having such small differences that drivers would not tend to notice or would quickly forget that there is any difference. Driveability should be acceptable in systems that have been set up and maintained properly. Cold-start can be problematic since the mechanical-control systems rely on engine airflow to meter propane. Less than precise metering at these low airflow rates can easily result in a mixture that is too rich or too lean. (Gasoline fuel systems have the advantage of being able to inject the amount of fuel needed, and metering does not have to be as precise because only a small portion of the gasoline vaporizes in cold weather. Propane should theoretically have a cold-start advantage over gasoline since it vaporizes so quickly, even at low temperatures, but because of this propensity to vaporize, metering becomes more important.)

Data are not widely available, but it is likely that liquid propane fuel systems should have improved vehicle acceleration relative to vaporized propane fuel systems. Vaporization of the propane would occur right in the intake port, cooling the intake air and regaining some of the volumetric efficiency loss that fully vaporized propane systems experience. More precise control of propane metering, especially during acceleration, should also improve vehicle acceleration performance. Such propane fuel systems should also have excellent driveability and cold-start performance the same or better than gasoline vehicles.

Heavy-Duty Vehicles

Very few heavy-duty propane vehicles have been developed and put into use; therefore, a database of knowledge about their performance characteristics does not exist. However, their characteristics should be similar to the relative differences between natural gas heavy-duty vehicles and their diesel engine counterparts. If this supposition holds true, heavy-duty propane vehicles should have similar or better power, the same or better driveability, and better cold-start performance compared to the same vehicle with a diesel engine. (Unlike light-duty vehicles, heavy-duty propane vehicles should have better cold-start performance compared to diesel engines because of the many cold-start challenges diesel engines face.)

Vegetable Oils

The interest in using vegetable oils as alternative fuels originated within the agricultural community as a fuel for agricultural tractors and equipment. In 1982, the American Society of Agricultural Engineers held the International Conference on Plant and Vegetable Oils as Fuels [1.28]. This conference made the point that vegetable oils could be viable alternative fuels for use in diesel engines. About the same time, vegetable oils were being studied in Austria, France, Germany, and Italy [1.29]. In Europe, vegetable oils were tested in engine dynamometer, and field tests were conducted in tractors, trucks, and diesel engine passenger cars.

The most popular types of crops from which vegetable oils can be extracted include soybeans, sunflowers, peanuts, rapeseed, and Chinese tallow trees. Dozens of candidate plants can yield significant oil yields per acre [1.30].

Initially, it was believed that vegetable oils could be used directly with minimal processing and preparation. However, extensive engine testing proved that while diesel engines operated satisfactorily on "raw" vegetable oils, combustion residues and deposits would quickly cause problems with fuel injectors, piston rings, and oil stability. By reacting the oils with methanol or ethanol, esters are formed which have much improved characteristics as fuels. These esterified versions of vegetable oils have been given the generic label of "biodiesel." Manufacturers of biodiesel have been targeting transit bus fleets to use a blend typically containing 20% biodiesel with diesel fuel, known also as "B20."[12] The favorable emissions properties of biodiesel reduce smoke, particulates, and gaseous emissions when used in the typical transit bus. The major impediment to use of such blends of biodiesel and diesel fuel is the cost of biodiesel compared to some other alternative fuels that could be used, and with emission control hardware for older transit bus engines.[13] Biodiesel is not the only alternative fuel facing unfavorable economics relative to petroleum fuels; however, it remains a viable contender as an alternative fuel for diesel transportation engines.

---

[12] These blends of biodiesel and diesel fuel are also frequently referred to as biodiesel. This book will use biodiesel to refer to 100% biodiesel while blends with diesel fuel will be referred to by their blend percentage, for example, 20% biodiesel would be called B20.

[13] The EPA finalized regulations, requiring all transit buses to install emissions control hardware when their engines are rebuilt, on April 21, 1993, to take effect January 1, 1995, in all metropolitan areas with 1980 populations of 750,000 or more.

The initial interest in using biodiesel was based on energy security for the agricultural segment. This interest remains, but there are several favorable environmental reasons for using biodiesel. Besides lower gaseous emissions, biodiesel contains no toxins (other than perhaps very small amounts of unreacted methanol in methyl ester biodiesels) and quickly biodegrades when spilled on the ground or in the water. An example of this interest in Europe is that chainsaw lubrication oils must be 80% biodegradable, which biodiesel satisfies readily.

Production

It is predicted that eventually biodiesel may reach 5-7% market share in Europe for diesel fuels [1.29]. This estimate is based on current and planned construction of biodiesel production facilities. When all these facilities are in place, it is estimated that over 1 million tons of biodiesel will be produced in the European Union, with 40,000 additional tons in the Czech Republic.

Vehicle Emissions Characteristics

Early testing of vegetable oils and biodiesel focused on engine power, reliability, and durability; emissions were rarely measured. Those proposing to use biodiesel in blends with diesel fuel have performed some emissions tests, and from these tests some generalizations about the emissions characteristics of biodiesel can be made [1.31].

Straight biodiesel (soy methyl ester) has a cetane rating significantly higher than typical No. 2 diesel fuel, slightly lower heating value, slightly higher viscosity, and contains approximately 10 mass percent oxygen. The lower heating value will cause a small loss in maximum power if the engine fuel system is not recalibrated. In a prechamber diesel engine using a transient eight-mode test, straight soy methyl ester showed a significant reduction in hydrocarbon emissions, no significant change in carbon monoxide emissions, a slight reduction in oxides of nitrogen emissions, reduced particulate emissions, and lower mutagenicity of the particulate matter formed. These results may be less favorable if the engine were recalibrated to the same maximum power output as when using No. 2 diesel fuel. Additional benefits of using soy methyl ester include reduced toxic emissions, very low sulfate emissions, and a much more pleasant exhaust odor. These beneficial emissions effects were attributed to the high cetane value of the soy methyl ester that caused reduced ignition delay, earlier energy release, and probably reduced peak flame temperatures.

When biodiesel is blended with diesel fuel, the emissions results change somewhat. A significant decrease in hydrocarbon and carbon monoxide emissions is typical, no change or a small increase in oxides of nitrogen emissions, and significant reduction in particulate emissions. Emissions of toxins would also decrease according to the percentage substitution of diesel fuel.

Biodiesel has inherently low sulfur content that makes it well-suited to diesel engines equipped with catalysts for emissions control. This fact and the fact that biodiesel in neat form does not have any aromatics, makes biodiesel competitive with "clean diesel" without any additional modifications.[14]

Vehicle Performance Characteristics

On a mass basis, neat biodiesel has approximately 13% less energy than typical diesel fuel. This loss in energy is caused by the oxygen content of biodiesel of approximately 10%. Biodiesel's higher specific gravity of approximately 0.88 compared to approximately 0.82 for diesel fuel regains some of the loss in energy on a mass basis for an overall impact of approximately 7% loss in energy per unit volume. Thus, an engine adjusted for diesel fuel should experience a loss of power of approximately 7% when using neat biodiesel (blends should experience power losses proportionate to the blend level). Engines readjusted to increase fuel injection quantities at full power should not experience any loss in power.

Because of the lower energy per unit volume, vehicles using neat biodiesel should experience a loss in fuel economy of about 7% on average.

Biodiesel has higher viscosity and higher pour points compared to typical diesel fuel, which could affect operation in very cold temperatures. Like diesel fuels, pour point additives are effective at decreasing pour point.

---

[14] "Clean diesel" is a term that applies to diesel fuel that has low sulfur and aromatic content, along with other characteristics that facilitate low emissions from diesel engines.

Engine oil dilution is a potential problem with biodiesel since it is more prone to oxidation and polymerization than diesel fuel. The presence of biodiesel in engine oil could cause thick sludge to occur with the consequence that the oil becomes too thick to pump. Special formulations of engine oil are being developed to minimize the effects of dilution with biodiesel.

Hydrogen

Hydrogen has many characteristics that make it the "ultimate" alternative fuel to fossil energy fuels. Hydrogen can be combusted directly in internal-combustion engines or it can be used in fuel cells to produce electricity with high efficiency (30-50% over the typical load range). When hydrogen is oxidized in fuel cells, the only emission is water vapor. When hydrogen is combusted in internal-combustion engines, water vapor is again the major emission, though some oxides of nitrogen may be formed if combustion temperatures are high enough. Therefore, the use of hydrogen as a transportation vehicle fuel would result in few or no emissions that would contribute to ozone formation.

Hydrogen is now produced primarily from the steam reformation of natural gas, though it can be produced from almost any source containing hydrogen in its composition. Hydrogen can also be produced from the electrolysis of water. This production route is desirable from an air quality standpoint only if the electricity is made from sources that do not use fossil fuels such as hydropower or nuclear energy. In the early 1970s, there was a widespread belief in the scientific community that the U.S. would move to the "all-electric economy" using electricity produced first from fossil fuels, then nuclear energy, and then solar energy [1.32]. Hydrogen would be the "energy carrier" in this system, distributed using the existing pipeline system created for natural gas. The hydrogen would be used in place of all the fuels currently used for stationary power and heating, and as a transportation fuel. Such a system would be very clean and would not produce by-products that contribute to global warming as presently is the case. The recent concern about nuclear waste and the cessation of nuclear power plant building has altered the conceptual path to the "hydrogen economy." Research is underway to develop novel, non-polluting means of hydrogen production such as from algae that makes use of sunlight or other biological methods.

The major drawback to using hydrogen as a fuel is the storage medium. Compared to all other fuels, hydrogen has the lowest energy storage density. Hydrogen can be stored as a compressed gas at pressures similar to CNG, liquefied, or stored in metal hydrides (which absorb hydrogen when cool and release it when heated) or carbon absorbents. Development of a practical fuel cell for vehicle applications would do much to promote the use of hydrogen as a fuel since the high efficiency of the fuel cell would reduce the storage requirements of hydrogen onboard the vehicle.

Production

In 1995, the U.S. produced 6.7 trillion liters (238 billion cu. ft.) of hydrogen, though this amount does not include hydrogen produced in refineries used for hydrotreating petroleum products [1.33]. This amount of hydrogen represents a little less than two days of average gasoline consumption in the U.S. in 1995 [1.34]. To use hydrogen as a vehicle fuel will most likely require it to be compressed to high pressure or liquefied. (Absorbent storage is another possibility but this technology has yet to achieve the energy storage density of compression or liquefaction.) In either case, a substantial portion of the energy of the hydrogen is consumed in compression or liquefaction.

Vehicle Emissions Characteristics

When oxidized in a fuel cell, the only significant emission is water vapor. When combusted in an internal-combustion engine (spark-ignition or diesel) some oxides of nitrogen and peroxides may be produced, depending on the calibration of the fuel system and configuration of the engine. None of the toxic emissions typical of petroleum fuels are present [1.35].

Like CNG or propane vehicles, hydrogen vehicles should not produce evaporative emissions since the fuel system would be closed. Even if hydrogen is released (e.g., fuel spills or vehicle maintenance) it rises quickly (being lighter than air) and does not cause any reactions in the atmosphere.

Vehicle Performance Characteristics

Only experimental hydrogen vehicles have been built to date, and it is not possible to derive meaningful conclusions about vehicle performance characteristics from them. Based on the internal-combustion engine work conducted to date, hydrogen engines should be able to produce the same amount of power that petroleum fuel engines do with superior efficiency since the lean limit of hydrogen is much lower than petroleum or other alternative fuels. A major concern about hydrogen vehicles will be operating range. A highly efficient drivetrain combined with a vehicle design that has low aerodynamic drag and rolling resistance may be necessary to achieve practical vehicle operating range.

Sources of Additional Information

Information about alternative fuels in addition to that presented in this chapter can be obtained from:

- U.S. Department of Energy
  Alternative Fuel Transportation Program
  Office of Energy Efficiency and Renewable Energy, EE-33
  1000 Independence Avenue, S.W.
  Washington, D.C. 20585

- The Department of Energy provides the Energy Efficiency and Renewable Energy Clearinghouse (EREC) that provides information via phone, mail, fax, bulletin board, and the internet.

  Phone: 1-800-DOE-EREC (363-3732)
  TDD (Telecommunications Device for the Deaf): 1-800-273-2957
  Computer Bulletin Board: 1-800-273-2955
  Fax: 1-703-893-0400
  Internet Address: http://www.eren.doe.gov
  Internet Electronic Mail: energyinfo@delphi.com

- The Department of Energy also provides information about alternative fuels through the National Alternative Fuels Hotline and Data Center. This service is available via phone and the internet, and provides assistance in locating specific information and reports about alternative fuels.

  Phone: 1-800-423-1DOE
  Internet Address: http://www.afdc.doe.gov

- Several states have conducted numerous studies and demonstrations of alternative fuel vehicles and can provide information and assistance. The following are among the most active:

The California Energy Commission
Transportation Technology & Fuels
1516 Ninth Street, MS-41
Sacramento, California 95814
Phone: 916-654-4634
Internet Address: http://www.energy.ca.gov

The New York State Energy Research and Development Authority
Corporate Plaza West
286 Washington Avenue Extension
Albany, New York 12203-6399
Phone: 518-862-1090
Fax: 518-862-1091
Internet Address: http://www.nyserda.org

Texas Department of Transportation
Dewitt C. Greer State Highway Building
125 East 11th Street
Austin, Texas 78701-2483
Phone: 512-463-8585

- The following organizations have information about alternative fuel vehicles:

National Association of Fleet Administrators
Islelin, New Jersey 08830
Phone: 908-494-8100

American Trucking Associations, Inc.
2200 Mill Road
Alexandria, Virginia 22314-4677
Phone: 703-838-1700
Internet Address: http://www.trucking.org

- Several newsletters provide news and information about alternative fuels. The following are two that provide information about all alternative fuels:

The Clean Fuels Report
J.E. Sinor Consultants, Inc.
6964 North 79th Street, Suite 1
P.O. Box 649
Niwot, Colorado 80544
Phone: 303-652-2632
Fax: 303-652-2772

Hart's 21st Century Fuels
Hart/IRI Fuels Information Services
1925 North Lynn Street
Arlington, Virginia 22209
Phone: 703-528-2500
Fax: 703-528-1603

- For more information about electric vehicles, the following associations can be contacted:

Electric Transportation Coalition
701 Pennsylvania Avenue, N.W., 4th Floor
Washington, D.C. 20004
Phone: 202-508-5995

Electric Vehicle Association of the Americas
601 California Street, Suite 502
San Francisco, California 94108
Phone: 408-253-5262

Edison Electric Institute
701 Pennsylvania Avenue, N.W., 4th Floor
Washington, D.C. 20004
Phone: 202-508-5000

Electric Power Research Institute
P.O. Box 10412
Palo Alto, California 94303
Phone: 415-855-2984

Electric Vehicle Industry Association
P.O. Box 85905
Tucson, Arizona 85754
Phone: 602-889-0248

- For more information about methanol, methanol vehicles, and methanol refueling infrastructure, contact:

The American Methanol Institute
800 Connecticut Avenue, N.W., Suite 620
Washington, D.C. 20006
Phone: 202-467-5050

American Automobile Manufacturers Association
7430 Second Avenue, Suite 300
Detroit, Michigan 48202
Phone: 313-872-4311
Fax: 313-872-5400

- For more information about ethanol, ethanol vehicles, and ethanol refueling infrastructure, contact:

Renewable Fuels Association
One Massachusetts Avenue, N.W.
Washington, D.C. 20001
Phone: 202-289-3835
Fax: 202-289-7519
Internet Address: http://www.EthanolRFA.org

American Automobile Manufacturers Association
7430 Second Avenue, Suite 300
Detroit, Michigan 48202
Phone: 313-872-4311
Fax: 313-872-5400

Clean Fuels Development Coalition
7315 Wisconsin Ave., East Tower, Suite 515
Bethesda, Maryland 20814
Phone: 301-913-3633

American Biofuels Association
1925 North Lynn Street, Suite 1050
Arlington, Virginia 22209
Phone: 703-522-3392

National Corn Growers Association
1000 Executive Parkway, Suite 105
St. Louis, Missouri 63141
Phone: 314-275-9915

Governors' Ethanol Coalition
P.O. Box 95085
Lincoln, Nebraska 68509-5085
402-471-2941

- For more information about natural gas, natural gas vehicles, and natural gas refueling infrastructure, contact:

Natural Gas Vehicle Coalition
1515 Wilson Boulevard, Suite 1030
Arlington, Virginia 22209
Phone: 703-527-3022
Fax: 703-527-3025

American Gas Association
1515 Wilson Boulevard, Suite 1030
Arlington, Virginia 22209
Phone: 703-841-8000

Gas Research Institute
8600 West Bryn Mawr Avenue
Chicago, Illinois 60631-3562
Phone: 312-399-8100
Fax: 312-399-8170
Internet Address: http://www.gri.org

- For more information about propane, propane vehicles, and propane refueling infrastructure, contact:

National Propane Gas Association
1600 Eisenhower Lane
Lisle, Illinois 60532
Phone: 708-515-0600
Fax: 708-515-8774

Propane Vehicle Council
1101 17th Street, N.W., Suite 1004
Washington, D.C. 20036
Phone: 202-530-0479
Fax: 202-466-7205

References

1.1. U.S. Environmental Protection Agency, "Environmental Fact Sheet—Tighter Controls Evaluated for NO_x, HC and PM Emissions from Heavy-Duty Engines," EPA420-F-95-008, June 1995.

1.2. Nash, A.W., and Hawes, D.A., *Principles of Motor Fuel Preparation and Application*, J.W. Wiley, 1938.

1.3. U.S. Department of Energy, "Assessment of Costs and Benefits of Flexible and Alternative Fuel Use in the U.S. Transportation Sector Technical Progress Report One: Context and Analytical Framework," DOE/PE-0080, January 1988.

1.4. Dr. Roberta Nichols, Manager of the Alternative Fuels Department of Ford's Automotive Emission and Fuel Economy Office, *Oxy-Fuel News*, November 5, 1990, pp. 3-4.

1.5. Douthit, W.H., "Effects of Oxygenates and Fuel Volatility on Vehicle Emissions at Seasonal Temperatures," SAE Paper No. 902180, Society of Automotive Engineers, Warrendale, Pa., 1990.

1.6. Ecklund, E.E., "Options for and Recent Trends in Use of Alternative Transportation Fuels," United Nations Centre for Human Settlements Ad Hoc Expert-Group Meeting in Human Settlements, 1986.

1.7. "Replacing Gasoline Alternative Fuels for Light-Duty Vehicles," Washington: Congress of the United States Office of Technology Assessment, 1990, p. 13.

1.8. U.S. Department of Energy, "Assessment of Costs and Benefits of Flexible and Alternative Fuel Use in the U.S. Transportation Sector, Technical Report Three: Methanol Production and Transportation Costs," November 1989.

1.9. "Growth of Top 50 Chemicals Slowed in 1995 from Very High 1994 Rate," *Chemical and Engineering News*, Vol. 74, No. 15, April 8, 1996.

1.10. Health Effects Institute, "HEI Strategic Plan for Vehicle Emissions and Fuels, 1994-1998," August 1994.

1.11. U.S. Environmental Protection Agency, "Mobil Source-Related Air Topics Study," as reported in *The Clean Fuels Report*, September 1993, pp. 11-14.

1.12. As reported in *The Clean Fuels Report*, Vol. 3, No. 3, June 1991, pp. 75-77, 96-97.

1.13. Burns, V.R., *et al.*, "Emissions with Reformulated Gasoline and Methanol Blends in 1992 and 1993 Model Year Vehicles," SAE Paper No. 941969, Society of Automotive Engineers, Warrendale, Pa., 1994.

1.14. Cowart, J.S., *et al.*, "Powertrain Development of the 1996 Ford Flexible Fuel Taurus," SAE Paper No. 952751, Society of Automotive Engineers, Warrendale, Pa., 1995.

1.15. "Assessment of Costs and Benefits of Flexible and Alternative Fuel Use in the U.S. Transportation Sector, Technical Report Seven: Environmental, Health, and Safety Concerns," DOE/PE-0100P, 1991, p. ix.

1.16. U.S. Department of Energy, "Alternatives to Traditional Transportation Fuels—1994—Volume 2: Greenhouse Gas Emissions," Energy Information Administration, Report No. DOE/EIA-0585(94)/1, February 1996.

1.17. Wang, M.Q., "Development and Use of the GREET Model to Estimate Fuel-Cycle Energy Use and Emissions of Various Transportation Technologies and Fuels," Argonne National Laboratory, Argonne, Ill.

1.18. "Accelerating the Market Penetration of Fuel Ethanol," prepared for the Congressional Research Service by Information Resources, Inc., April 15, 1992.

1.19. Hyunok Lee, "Ethanol's Evolving Role in the U.S. Automobile Fuel Market," *Industrial Uses of Agricultural Materials*, U.S. Department of Agriculture, Office of Energy, June 1993.

1.20. Dodge, L.G., *et al.*, "Development of a Dedicated Ethanol Ultra-Low Emission Vehicle (ULEV)—Phase 3 Report," National Renewable Energy Laboratory, Golden, Co., September 1996.

1.21. Benson, J.D., *et al.*, "Emissions with E85 and Gasolines in Flexible/Variable Fuel Vehicles—The Auto/Oil Air Quality Improvement Research Program," SAE Paper No. 952508, Society of Automotive Engineers, Warrendale, Pa., 1995.

1.22. U.S. Department of Energy, "Monthly Energy Review—October 1995," DOE/EIA-0035(95/10), Energy Information Administration, Office of Energy Markets and End Use.

1.23. Oppenheimer, Ernest J., Ph.D., *Natural Gas: The New Energy Leader*, Pen and Podium Productions, New York, N.Y., July 1981.

1.24. Disbrow, Jim, "Data Collection on Alternative-Fuel Vehicles," *Monthly Energy Review*, Energy Information Administration, October 1994.

1.25. U.S. Department of Energy, "Assessment of Costs and Benefits of Flexible and Alternative Fuel Use in the U.S. Transportation Sector—Progress Report One: Context and Analytical Framework," January 1988.

1.26. Bass, E., Bailey, B., and Jaeger, S., "LPG Conversion and HC Emissions Speciation of a Light-Duty Vehicle," SAE Paper No. 932745, Society of Automotive Engineers, Warrendale, Pa., 1993.

1.27. *Federal Register*, Vol. 59, No. 182, September 21, 1994, pp. 43472-48538.

1.28. *Vegetable Oil Fuels*, Proceedings of the International Conference on Plant and Vegetable Oils as Fuels, August 2-4, 1982, American Society of Agricultural Engineers, St. Joseph, Mich.

1.29. Korbitz, Werner, "Status and Development of Biodiesel Production and Projects in Europe and North America," SAE Paper No. 952768, Society of Automotive Engineers, Warrendale, Pa., 1995.

1.30. Duke, James A., and Bagby, Marvin O., "Comparison of Oilseed Yields: A Preliminary Review," Proceedings of the International Conference on Plant and Vegetable Oils as Fuels, August 2-4, 1982, American Society of Agricultural Engineers, St. Joseph, Mich.

1.31. McDonald, J., Purcell, D.L., McClure, B.T., and Kittelson, D.B., "Emissions Characteristics of Soy Methyl Ester Fuels in an IDI Compression Ignition Engine," SAE Paper 950400, Society of Automotive Engineers, Warrendale, Pa., 1995.

1.32. Gregory, Derek P., "The Hydrogen Economy," *Scientific American*, Vol. 228, No. 1, January 1973.

1.33. "Facts & Figures for the U.S. Chemical Industry," *Chemical and Engineering News*, Vol. 74, No. 26, June 24, 1996.

1.34. U.S. Department of Energy, "Monthly Energy Review—July 1996," DOE/EIA-0035(96/07), Energy Information Administration, Office of Energy Markets and End Use.

1.35. Swain, M.R., Adt, R.R., Jr., and Pappas, J.M., "Experimental Hydrogen-Fueled Automotive Engine Design Data-Base Project," U.S. Department of Energy Report DOE/CS/51212-1, Vol. 3, May 1983.

1.36. Lapetz, John, *et al.*, "Ford's 1996 Crown Victoria Dedicated Natural Gas Vehicle," SAE Paper No. 952743, Society of Automotive Engineers, Warrendale, Pa., 1995.

Chapter Two

Properties and Specifications

The properties of a fuel define its physical and chemical characteristics. A thorough understanding of the properties of a fuel is essential to the design and engineering of engine combustion systems, vehicle fuel systems, fuel storage, and fuel dispensing systems. Fuel properties impact vehicle performance, emissions, fuel efficiency, reliability, and durability.

Fuel properties are especially important in defining the safety hazards posed by a fuel. Since fuels are flammable, fire and explosion hazards are possible. Some fuels are toxic or contain carcinogenic compounds that present exposure, inhalation, and ingestion hazards. Fuels stored at cryogenic temperatures such as liquefied natural gas and liquefied hydrogen present safety hazards from skin contact.

In a more perfect world, fuel properties would be the same from all producers everywhere. Unfortunately, this is not the case, and fuel properties vary according to production plant, feedstock variations and quality, production quality control, and the presence of contaminants picked up during storage and distribution.

Fuel specifications represent an attempt to mold and limit fuel properties to facilitate use in vehicles and limit the hazards presented in storing and handling fuels. Petroleum fuels have an advantage here in that producers have some latitude to vary the properties of the final product. There is no such option for some fuels such as natural gas which is predominately methane, and ideally would be 100% methane. Methanol and ethanol are also single-constituent fuels, but it is possible to vary their properties advantageously through the addition of gasoline or other additives.

Fuel specifications also serve a very valuable service as a means of standardizing fuel properties, which facilitates trade and commerce of that fuel. For example, if each producer were allowed to produce fuels that varied in properties, the chances of vehicle manufacturers producing vehicles with acceptable and uniform performance, driveability, and fuel efficiency would be next to impossible. The fuel distribution system is based on the concept of fuel fungibility, i.e., the ability to trade fuels within a wide region without fear of problems due to variations in fuel properties.

Besides limits on physical and chemical properties, fuel specifications include tests that evaluate the performance characteristics of a fuel. For example, there are tests to assess the corrosivity of fuels on fuel system materials and performance tests to evaluate vapor pressure, octane number, cetane number, and others. In many cases, additives are used to assist fuels in satisfying the performance test portions of fuel specifications.

As just discussed, fuel properties affect many aspects of vehicle design, fuel storage and distribution, and safety hazards. Following are definitions of the fuel properties of most interest and an explanation of their significance. Later in this chapter, properties and specifications for the alternative fuels will be presented and compared to typical gasoline and diesel fuel.

Autoignition Temperature—Minimum temperature of a substance to initiate self-sustained combustion independent of any ignition source [2.1].

Boiling Temperature—Temperature at which the transformation from liquid to vapor phase occurs in a substance at a pressure of 14.7 psi (atmospheric pressure at sea level). Fuels that are pure compounds (such as methanol) have a single temperature as their boiling points, while fuels with mixtures of several compounds (like gasoline) have a boiling temperature range representing the boiling points of each individual compound in the mixture. For these mixtures, the 10% point of distillation is often used as the boiling point [2.1].

Cetane Number—The ignition quality of a diesel fuel measured using an engine test specified in ASTM Method D 613. Cetane number is determined using two pure hydrocarbon reference fuels: cetane, which has a cetane rating of 100; and heptamethylnonane (also called isocetane), which has a cetane rating of 15 [2.2].

Density—Mass per unit volume, expressed in kg/l or lb/gal [2.3].

Electrical Conductivity—Measure of the ability of a substance to conduct an electrical charge.

Flame Spread Rate—Rate of flame propagation across a fuel pool.

Flame Visibility—Degree to which combustion of a substance under various conditions can be seen.

Flammability Limits—Minimum and maximum concentrations of vapor in air below and above which the mixtures are unignitable [2.1, 2.3]. A vapor-air concentration below the lower flammable limit is too lean to ignite, while a concentration above the upper flammable limit is too rich to ignite.

Flash Point—Minimum temperature of a liquid at which sufficient vapor is produced to form a flammable mixture with air [2.1, 2.3].

Freezing Point—The temperature where a liquid can exist as both a liquid and a solid in equilibrium [2.3].

Heating Value—The heat released when a fuel is combusted completely, corrected to standard pressure and temperature. The higher heating value is complete combustion with the water vapor in the exhaust gases condensed. The lower heating value is when the water vapor in the exhaust is in the vapor phase [2.4].

Latent Heat of Vaporization—The quantity of heat absorbed by a fuel in passing between liquid and gaseous phases [2.1]. The conditions under which latent heat of vaporization is measured is the boiling point (or boiling range) and atmospheric pressure, 101.4 kPa (14.7 psi).

Molecular Weight—The sum of the atomic weights of all the atoms in a molecule [2.1].

Octane Number—A measure of the resistance of a fuel to combustion knock using standardized engine tests. The research octane number is determined using

ASTM Method D 2699; the motor octane number is determined using ASTM Method D 2700. The Antiknock Index is the average of the Research and Motor numbers. Octane numbers are determined using n-heptane that has an octane number of 0, and isooctane that has an octane number of 100 [2.2].

Odor Recognition—Degree of smell associated with fuel vapor.

Specific Gravity—The ratio of the density of a material to the density of water [2.3].

Specific Heat—The ratio of the heat needed to raise the temperature of a substance one degree compared to the heat needed to raise the same mass of water one degree [2.3].

Stoichiometric Air-fuel Ratio—The exact air-fuel ratio required to completely combust a fuel to water and carbon dioxide [2.4].

Vapor Density—Weight of a volume of pure (no air present) vapor compared to the weight of an equal volume of dry air at the same temperature and pressure. A vapor density of less than one describes a vapor which is lighter than air, while a value greater than one describes a vapor that is heavier than air [2.1].

Vapor Pressure—Equilibrium pressure exerted by vapors over a liquid at a given temperature [2.1, 2.3]. The Reid vapor pressure (RVP) is typically used to describe the vapor pressure of petroleum fuels without oxygenates at 100°F (ASTM Test Method D 323, Test Method for Vapor Pressure of Petroleum Products) [2.5]. The term "true vapor pressure" is often used to distinguish between vapor pressure and Reid vapor pressure. The Reid vapor pressure test involves saturating the fuel with water before testing and cannot be used for gasoline-alcohol blends or neat alcohol fuels; a new procedure has been developed which does not use water and is called Dry Vapor Pressure Equivalent, or DVPE (see ASTM D 4814-95c under Additional Information section).

Viscosity—The resistance of a liquid to flow [2.3].

Water Solubility—Maximum concentration of a substance that will dissolve in water.

The Alcohols

Methanol and ethanol are in many ways physically very similar to gasoline. They weigh about the same and can be stored using containers of the same design as gasoline. Beyond these most basic similarities, there are many differences between gasoline and alcohols, as explained in the following.

Methanol

Methanol is a colorless liquid with a faint odor. It looks like water, but just dipping a finger in methanol and removing it reveals one of methanol's signature characteristics: its high latent heat of vaporization cools your finger rapidly as it evaporates. (Since methanol is toxic and can be absorbed through the skin, prolonged skin exposure is not recommended.)

Referring to Table 2-1, the formula for methanol illustrates a composition that includes oxygen, unlike gasoline which does not include any oxygen unless added during the refining process. Methanol is one-half oxygen which plays an important role in its combustion characteristics. Both methanol and gasoline contain about the same amount of hydrogen, but methanol contains much less carbon due to the oxygen that is present.

Methanol is a single compound of much lighter weight compared to the several dozen, heavier hydrocarbon compounds that make up gasoline. However, the density of methanol is very similar to gasoline, and a liter of methanol weighs 30-90 g more than a liter of typical gasoline (2-10 oz more than a typical gallon of gasoline).

Methanol has a very low freezing point which is essential for a practical fuel, and indeed, methanol was once used as an engine coolant antifreeze before better antifreeze compounds were found. While methanol's freezing point is much lower than gasoline and diesel fuel, this is no particular advantage to methanol since gasoline's freezing point is sufficiently low to prevent problems in vehicle fuel systems.

At the other end of the liquid scale, methanol boils at one temperature (65°C [149°F]) unlike gasoline which boils over a very wide temperature range. This characteristic of methanol, along with its low vapor pressure (32 kPa @ 38°C

Table 2-1. Methanol Compared to Gasoline and Diesel Fuel

Fuel Property	Methanol[a]	Gasoline[a]	No. 2 Diesel Fuel[a]
Formula	CH_3OH	C_4 to C_{12}	C_8 to C_{25}
Molecular Weight	32.04	100-105	200 (approx.)
Composition, Weight %			
Carbon	37.5	85-88	84-87
Hydrogen	12.6	12-15	13-16
Oxygen (for gasolines, oxygenated or reformulated gasolines)	49.9	0-4	0
Density, kg/L, 15/15°C	0.796	0.69-0.79	0.81-0.89
(lb/gal), 60/60°F	(6.63)	(5.8-6.6)	(6.7-7.4)
Specific Gravity (Relative Density), 15/15°C	0.796	0.69-0.79	0.81-0.89
Freezing Point, °C	−97.5	−40	−40 to −1
(°F)	(−143.5)	(−40)	(−40 to 30)
Boiling Point, °C	65	27-225	188-343
(°F)	(149)	(80-437)	(370-650)
Vapor Pressure, kPa @ 38°C	32	48-103	<1
(psi @ 100°F)	(4.6)	(7-15)	(<0.2)
Specific Heat, kJ/(kg-K)	2.5	2.0	1.8
(Btu/lb-°F)	(0.60)	(0.48)	(0.43)
Viscosity, mPa-s @ 20°C	0.59	0.37-0.44	2.6-4.1
(mm/s @ 68°F)	(0.74)	(0.5-0.6)	(2.8-5.0)
Water Solubility, 21°C (70°F)			
Water in Fuel, Vol %	100	Negligible	Negligible
Electrical Conductivity, mhos/cm	4.4×10^{-7}	1×10^{-14}	1×10^{-12}
Latent Heat of Vaporization,			
kJ/kg	1178	349	233
(Btu/lb)	(506)	(150)	(100)
Lower Heating Value, 1000 kJ/L	15.8	30-33	35-37
(1000 Btu/gal)	(56.8)	(109-119)	(126-131)
Flash Point, °C	11	−43	74
(°F)	(52)	(−45)	(165)
Autoignition Temperature, °C	464	257	316
(°F)	(867)	(495)	(600)
Flammability Limits, Vol %			
Lower	7.3	1.4	1.0
Higher	36.0	7.6	6.0
Stoichiometric Air-Fuel Ratio, Weight	6.45	14.7	14.7

Table 2-1. Methanol Compared to Gasoline and Diesel Fuel *(cont.)*

Fuel Property	Methanol[a]	Gasoline[a]	No. 2 Diesel Fuel[a]
Flame Spread Rate, m/s	2-4[b]	4-6[b]	-
(ft/sec)	(7-13)	(13-20)	-
Flame Visibility	Invisible in Daylight[c]	Visible in all Conditions[b]	Visible in all Conditions[b]
Octane Number			
Research	108.7[d]	88-100	-
Motor	88.6	80-90	-
Cetane Number	-	-	40-55

[a] All properties are for conventional, non-oxygenated gasoline and diesel fuel from Ref. [2.17], except where noted. (Numbers include an update in progress as of February 1997.)
[b] Ref. [2.18]
[c] Ref. [2.19]
[d] Ref. [2.8]

[4.6 psi @ 100°F]), highlights a significant difference between methanol and gasoline—methanol is much less prone to produce flammable vapor at ambient temperatures. This difference is vividly illustrated by the fact that pure methanol will not start in fuel systems designed for gasoline at temperatures below about 7°C (45°F). This is not to say that methanol is relatively more safe when exposed to the atmosphere at room temperature. Rather, it does illustrate that the ignition source must have enough energy to vaporize sufficient liquid methanol to create a flammable mixture of methanol vapor and air.

Methanol's viscosity is a little higher than gasoline, but much lower than diesel fuel. When used as a diesel engine fuel, methanol's low viscosity has been identified as a reason methanol compression-ignition engines start much more quickly than compression-ignition engines using diesel fuel.

Methanol is completely soluble in water unlike gasoline and diesel fuel which are essentially non-soluble in water. (Actually, gasoline and diesel fuel will absorb a very small amount of water, depending on the amount of aromatics present.) This could prove to be a problem in commerce since the current fuel distribution and storage systems for gasoline and diesel fuel are not water-tight. While water that enters a gasoline or diesel fuel system is a nuisance, it does not immediately create fuel quality problems as it would with methanol. Also, a water-soluble fuel will likely be very tempting to those who are less than honorable in that water can be added at very low cost and sold at a fuel price. Less than 10% water would not significantly affect the combustion characteristics of methanol, though it would significantly affect the profitability of selling the fuel and possibly the long-term durability of the engines using it. The electrical conductivity of methanol is much higher than for gasoline or diesel fuel. This represents a significant safety advantage since static electric charge induced from fuel movement into and out of fuel tanks will dissipate much more quickly. The danger from static electrical charge decreases as the stored charge decreases. However, the high conductivity of methanol can cause severe galvanic or electrolytic corrosion in fuel systems designed for gasoline and diesel fuel which could have adverse safety implications.

Methanol has a latent heat of vaporization nearly 3.4 times that of gasoline on a mass basis. Since it takes approximately twice as much methanol as gasoline, the overall effect is nearly seven times as high. This is an advantage to the power output of spark-ignition engines because the latent heat cools the intake air, making it more dense. It also decreases the propensity for autoignition by cooling combustion temperatures, which also results in lower emissions of oxides of nitrogen, one of the major contributors to ozone formation. Methanol's high latent heat of vaporization is a safety concern in cold weather since skin contact can quickly cause frostbite.

Methanol does not produce vapors as easily as gasoline because gasoline contains many hydrocarbons with very low boiling points. The flash point of methanol is 11°C (52°F) compared to gasoline at –43°C (–45°F). This corresponds very well with actual test data where gasoline engines modified to use methanol found that starting would occur only at temperatures above 7°C (45°F). A high flash point is a safety advantage though in the "real world" what counts most is the amount of energy available to heat and vaporize the fuel. This is illustrated by the trick whereby lighted matches or cigarettes are thrust into liquid fuel with the only consequence that they are snuffed out. The result would be very different if the lighted match or cigarette were brought to the surface of the liquid fuel in a slow manner so that the heat could warm a small portion of the fuel and create sufficient vapor to allow ignition to occur. Once a small ignition occurs, it spreads very rapidly across the entire liquid surface.

The autoignition temperature of methanol is nearly twice that of gasoline and represents another inherent safety advantage. A higher autoignition temperature means that methanol vapor can be heated to higher temperatures before it ignites compared to gasoline vapor.

The flammability limits of methanol are both higher and wider than those of gasoline. This means that more methanol vapor must be created before a flammable mixture is possible. In fact, as gasoline vapor and air mixtures pass the rich limit (7.6%), methanol vapor and air mixtures are just becoming flammable (7.3%). Methanol vapor and air mixtures remain flammable until the methanol vapor increases beyond 36%. Thus, if vapor is generated at similar rates, gasoline achieves flammability sooner but quickly passes beyond the rich limit. Methanol, on the other hand, achieves flammability much later and stays in the flammable range much longer.

Because of the oxygen content of methanol, its stoichiometric air-fuel ratio (6.45) is less than half that of gasoline (14.7). For the same amount of energy production, about twice as much methanol is needed than gasoline, but about the same amount of air is consumed.

The flame spread rate of methanol is about two-thirds that of gasoline. This measure applies primarily to how fast flame will traverse a pool of fuel, and the difference between methanol and gasoline is not really significant since both will be engulfed in flame before any actions could be taken to stop flame propagation.

A very serious safety deficiency of methanol is that it burns without a visible flame in daylight which makes fire detection and fire fighting difficult. The invisible flame is an indication of the soot-free combustion properties of methanol. A moderate safety advantage of methanol flames is that they have low radiant heat transfer, meaning that surfaces close to a methanol flame will not get as hot as compared to a luminous flame such as from gasoline.

Methanol has good octane properties compared to gasoline. With a research octane value of 108.6 and a modified motor octane value[1] of 88.6 ([R+M]/2 of 98.6),

---

[1] The standard motor octane test must be modified to include fuel heaters to enable methanol to be tested.

methanol has sufficient octane to allow engines to use higher compression ratios with the attendant benefits of improved power and efficiency. The average octane of methanol is at least 5 numbers higher than the highest octane unleaded premium gasoline available, and 12 numbers more than the typical unleaded regular grade of gasoline.

Ethanol

Ethanol for use as a fuel is produced via fermentation of agricultural crops. Corn is the preferred crop in the U.S., though Brazil produces ethanol from sugar cane. Producing ethanol this way is similar to producing alcoholic beverages with the exception that the ethanol is separated from the mix of ethanol and water using distillation. In the U.S. and most other countries, beverages containing ethanol are highly taxed. To prevent ethanol intended for the fuel market from being sold into the beverage market at a high profit, denaturants are added to it to make it poisonous and unappetizing for consumption. Ethanol sold for fuel use in the U.S. is denatured using 5 volume percent regular unleaded gasoline [2.6].

Pure ethanol is a clear liquid with a characteristic though faint odor. Like methanol, ethanol has a high latent heat of vaporization which can be demonstrated by dipping a finger in ethanol and letting it dry which causes rapid cooling. In cold weather, frostbite is a concern when handling ethanol because its latent heat can cause exposed skin to freeze. (Skin contact should be avoided with all fuels to prevent absorption of toxic compounds into the body.)

Table 2-2 lists the properties of pure ethanol compared to gasoline and No. 2 diesel fuel. Like methanol, ethanol contains oxygen, which gasoline does not unless it is added. However, ethanol contains only about half the oxygen that methanol does (26.6% compared to 49.9%). Both ethanol and methanol contain about the same percentage of hydrogen, with ethanol containing more carbon than methanol. Methanol represents a large change in properties compared to gasoline because of its oxygen content, and ethanol properties are often found between those of methanol and gasoline, as its composition would suggest.

Ethanol is a single compound, like methanol, and nearly one and one-half times as heavy as methanol but still much lighter than the typical hydrocarbons that make up gasoline. The specific gravity of ethanol is higher than gasoline (and

Table 2-2. Ethanol Properties Compared to Gasoline and Diesel Fuel

Fuel Property	Ethanol[a]	Gasoline[a]	No. 2 Diesel Fuel[a]
Formula	C_2H_5OH	C_4 to C_{12}	C_8 to C_{25}
Molecular Weight	46.07	100-105	200 (approx.)
Composition, Weight %			
Carbon	52.2	85-88	84-87
Hydrogen	13.1	12-15	13-16
Oxygen (for gasolines, oxygenated and reformulated gasolines)	34.7	0-4	0
Density, kg/L, 15/15°C	0.79	0.69-0.79	0.81-0.89
(lb/gal), 60/60°F	(6.61)	(5.8-6.6)	(6.7-7.4)
Specific Gravity (Relative Density) 15/15°C	0.794	0.69-0.79	0.81-0.89
Freezing Point, °C	−114	−40	−40 to −1
(°F)	(−173.2)	(−40)	(−40 to 30)
Boiling Point, °C	78	27-225	188-343
(°F)	(172)	(80-437)	(370-650)
Vapor Pressure, kPa @ 38°C	15.9	48-103	<1
(psi @ 100°F)	(2.3)	(7-15)	(<0.2)
Specific Heat, kJ/(kg-K)	2.4	2.0	1.8
(Btu/lb-°F)	(0.57)	(0.48)	(0.43)
Viscosity, mPa-s @ 20°C	1.19	0.37-0.44	2.6-4.1
(mm/s @ 68°F)	(1.50)	(0.5-0.6)	(2.8-5.0)
Water Solubility, 21°C (70°F)			
Water in Fuel, Vol %	100	Negligible	Negligible
Electrical Conductivity, mhos/cm	1.35×10^{-9}	1×10^{-14}	1×10^{-12}
Latent Heat of Vaporization, kJ/kg	923	349	233
(Btu/lb)	(396)	(150)	(100)
Lower Heating Value, 1000 kJ/L	21.1	30-33	35-37
(1000 Btu/gal)	(76)	(109-119)	(126-131)
Flash Point, °C	13	−43	74
(°F)	(55)	(−45)	(165)
Autoignition Temperature, °C	423	257	316
(°F)	(793)	(495)	(600)
Flammability Limits, Vol %			
Lower	4.3	1.4	1.0
Higher	19.0	7.6	6.0
Stoichiometric Air-Fuel Ratio, Weight	9.00	14.7	14.7

53

**Table 2-2. Ethanol Properties Compared to Gasoline
and Diesel Fuel (*cont.*)**

Fuel Property	Ethanol[a]	Gasoline[a]	No. 2 Diesel Fuel[a]
Flame Visibility	Difficult to see in Daylight	Visible in all Conditions[b]	Visible in all Conditions[b]
Octane Number			
Research	108.6[c]	88-100	-
Motor	89.7	80-90	-
Cetane Number		-	40-55

[a] All properties are for conventional, non-oxygenated gasoline and diesel fuel from Ref. [2.17], except where noted. (Numbers include an update in progress as of February, 1997.)
[b] Ref. [2.18]
[c] Ref. [2.8]

about the same as methanol). A liter of ethanol weighs 20-80 g more per liter than typical gasoline (2-10 oz. more than the typical gallon of gasoline).

Ethanol's freezing point is much lower than gasoline, though it rises quickly as the water content of ethanol increases. Freezing of ethanol in fuel storage and distribution systems is not a concern.

The boiling point of ethanol is higher than that of methanol (78°C compared to 65°C [172°F compared to 149°F]) and much higher than the initial boiling point of gasoline. Ethanol's high boiling point and low vapor pressure (15.9 kPa @ 38°C [2.3 psi @ 100°F]) makes formation of flammable ethanol vapors at ambient conditions less likely than for gasoline. Ethanol's viscosity is higher than gasoline but is still less than diesel fuel. No problems are known for handling of ethanol in cold weather due to high viscosity.

Ethanol is completely soluble in water which presents potential problems for storage and handling. Current fuel distribution and storage systems are not water-tight, and water tends to carry other impurities with it. Ethanol will not be significantly degraded by small amounts of clean water, though water addition dilutes its value as a fuel.

Ethanol's electrical conductivity is much higher than gasoline or diesel fuel, though lower than methanol. A higher conductivity suggests that ethanol will dissipate static charges that build up when pumping fuel during fuel transfers faster than gasoline. This gives ethanol a theoretical safety advantage over gasoline when static electrical charges are created in that they will be dissipated more quickly. But as with methanol, ethanol's high conductivity may lead to galvanic and electrolytic corrosion in fuel systems designed for gasoline and diesel fuel, which could have adverse safety implications.

The oxygen content of ethanol reduces its heating value relative to gasoline. Ethanol has between 64% and 70% of the energy per gallon that gasoline has (depending on the heating value of gasoline). To produce the same amount of power as gasoline, about 1.5 times the volume of ethanol is needed. Heating value is one area where ethanol has an advantage over methanol. It takes about 1.3 gallons of methanol to equal the energy of each gallon of ethanol.

The autoignition temperature of ethanol is significantly higher than that of gasoline or diesel fuel. This should make ethanol less susceptible to ignition when spilled on hot surfaces such as engine exhaust manifolds.

The flammability limits of ethanol are higher than gasoline and diesel fuel, but lower than methanol. This suggests that more ethanol (than gasoline) vapor would have to be produced and mixed with air before a flammable mixture is produced. Given ethanol's high latent heat of vaporization, low vapor pressure, and high boiling point, creating vapor requires much more energy than for gasoline. Once sufficient vapor is produced to make ethanol and air mixtures flammable, ethanol's upper flammability limit is higher than gasoline, representing a wider range in which flammable mixtures can exist. Which is less safe depends on the situation.

The oxygen content of ethanol lowers its stoichiometric air-fuel ratio relative to gasoline in proportion to the difference in heating value.

Pure ethanol burns with a flame that is difficult to see in bright sunlight, but it does not represent the same visibility difficulty that methanol does. When gasoline is added to ethanol, flame visibility is not a problem. Like methanol, the less luminous flame of ethanol has lower radiant heat transfer which should be a safety advantage when trying to contain an ethanol fire.

Ethanol has good octane properties compared to gasoline and about the same as methanol. With a research octane value of 108.6 and a motor octane value of 89.7 (R+M/2 of 99), ethanol has sufficient octane to allow engines to use higher compression ratios with the attendant benefits of improved power and efficiency. The average octane of ethanol is at least 5 numbers higher than the highest octane unleaded premium gasoline available, and 12 numbers more than the typical unleaded regular grade of gasoline.

M85 and E85

The low vapor pressure and high latent heat of vaporization of methanol and ethanol created cold-start difficulties in spark-ignition engines. To overcome this hurdle and improve flame visibility (especially for methanol), a consensus developed that 15 volume percent gasoline would be added to methanol and ethanol, known as M85 and E85, respectively. The addition of this much gasoline changes some of the fuel properties significantly and makes them behave much more like gasoline. M85 and E85 facilitated development of "flexible fuel vehicles" (FFVs) which—through the addition of a sensor that measures the percentage of alcohol in the fuel blend—allow straight gasoline and M85 or E85 to be used in the same fuel tank. Using M85 or E85, an FFV will cold-start under all but the coldest ambient temperatures when mixtures such as M75 and E75 are recommended.

Table 2-3 lists the properties of M85 and E85 compared to gasoline. The addition of gasoline increases the percentage of carbon in the fuel and decreases the percentage of oxygen, with hydrogen content remaining essentially unchanged. Density and specific gravity also are not changed significantly from the values for methanol and ethanol. The addition of gasoline to make M85 and E85 drastically changes the boiling characteristics of methanol and ethanol, which in pure form boil at a single temperature. Adding gasoline gives M85 and E85 initial boiling points approaching gasoline, with very flat sections at the boiling temperatures of methanol and ethanol. These flat sections do not appear to cause significant driveability or emissions problems for FFVs using these fuels. Gasoline addition also increases the final boiling points, though the current distillation test probably does not capture the true final boiling point of M85 or E85, which should approach that of gasoline.

The RVP of M85 and E85 tends to be compromised somewhat from that of gasoline. For example, if the RVP of the gasoline added is 76 kPa (11 psi), the RVP of

Table 2-3. M85 and E85 Properties Compared to Gasoline

Fuel Property	M85[a]	E85[a]	Gasoline[b]
Composition, Weight %			
Carbon	43-45	56-58	85-88
Hydrogen	12-13	13-14	12-15
Oxygen (for gasoline, oxygen comes from adding oxygenates)	43-44	29-30	0-4
Density, kg/L, 15/15°C	0.79-0.80	0.78-0.79	0.69-0.79
(lb/gal), 60/60°F	(6.5-6.6)	(6.5-6.6)	(5.8-6.6)
Specific Gravity (Relative Density), 15/15°C	0.79-0.80	0.79-0.80	0.69-0.79
Freezing Point, °C	-	-	–40
(°F)	-	-	(–40)
Boiling Point, °C	49-66[e]	49-80[e]	27-225
(°F)	(120-150)	(120-176)	(80-437)
Vapor Pressure, kPa @ 38°C	48-103	38-83	48-103
(psi @ 100°F)	(7-15)	(5.5-12)	(7-15)
Specific Heat, kJ/(kg-K)	2.4	2.3	2.0
(Btu/lb-°F)	(0.58)	(0.55)	(0.48)
Viscosity, mPa-s @ 20°C	0.55-0.57	1.07-1.08	0.37-0.44
(mm/s @ 68°F)	(-)	(-)	(0.5-0.6)
Water Solubility, 21°C (70°F) Water in Fuel, Vol %	100[d]	100[d]	Negligible
Electrical Conductivity, mhos/cm	-	-	1×10^{-14}
Latent Heat of Vaporization, kJ/kg	1055	836	349
(Btu/lb)	(453)	(359)	(150)
Lower Heating Value, 1000 kJ/L	17.9-18.3	22.4-22.9	30-33
(1000 Btu/gal)	(64.6-66.1)	(81.0-82.5)	(109-119)
Flash Point, °C	slightly warmer than gasoline	slightly warmer than gasoline	–43
(°F)			(–45)
Autoignition Temperature, °C	greater than gasoline	greater than gasoline	257
(°F)			(495)
Flammability Limits, Vol %			
Lower	wider than	between M85	1.4
Higher	gasoline	and gasoline	7.6
Stoichiometric Air-Fuel Ratio, Weight	7.7	9.9	14.7
Flame Spread Rate, m/s	slower than gasoline	slower than gasoline	4-6[c]
(ft/sec)			(13-20)

Table 2-3. M85 and E85 Properties Compared to Gasoline *(cont.)*

Fuel Property	M85[a]	E85[a]	Gasoline[b]
Flame Visibility	initially good; decreases	initially good; decreases	Visible in all Conditions[c]
Octane Number			
Research	108[f]	107[f]	88-100
Motor	89	89	80-90

[a] Properties for M85 and E85 extrapolated from M100, E100, and gasoline properties, except where noted.
[b] All properties are for conventional, non-oxygenated gasoline fuel from Ref.[2.17], except where noted. (Numbers include an update in progress as of February, 1997.)
[c] Ref. [2.18]
[d] The gasoline portion separates, but water will be 100% soluble in the alcohol portion.
[e] Ref. [2.20]
[f] Ref. [2.8]

M85 or E85 created using this gasoline will probably be between 62 kPa (9 psi) and 69 kPa (10 psi). To compensate, the gasoline used to produce M85 or E85 will contain additional butane and/or pentanes to counteract the damping effect of the low vapor pressure of methanol and ethanol.

The heating values of M85 and E85 also are improved by the addition of gasoline. It should take only 1.75 liters of M85 and 1.40 liters of E85 to equal a liter of gasoline. Some properties are degraded through the addition of gasoline. For instance, the flash point is lowered to some value very close to gasoline and the autoignition temperature is lowered.

Flame visibility is improved for both methanol and ethanol, but the impact varies with burning time. Studies of M85 pool fires have demonstrated that when an M85 fire is started, the flame color is very visible due to the gasoline present. But, as the flame burns over time, the gasoline is preferentially consumed, leaving behind a fuel that is essentially methanol. Over time, the flame visibility gets weaker until near the end when it is similar to pure methanol [2.7]. Certainly a good initial warning is better than a good final warning, however.

Little work has been done to measure the octane value of M85 and E85, but what has been done has shown their octane values to be only slightly degraded from pure methanol and ethanol [2.8]. The American Society for Testing and Materials (ASTM) has developed specifications (D 5797 and D 5798) for fuel methanol and fuel ethanol [2.9, 2.10]. Both specifications define three classes of fuel that vary in gasoline content and vapor pressure (see Table 2-4). The classes are designed to ensure that methanol and ethanol fuels have sufficient vapor pressure to achieve acceptable cold starts in all areas of the country during all months of the year. The specifications recommend which fuel classes should be used throughout the country around the year. While gasoline varies in vapor pressure throughout the year, these alcohol fuels would vary in both gasoline content and vapor pressure. An excellent discussion of the significance of M85 fuel specifications (with implications for E85) is available from the Society of Automotive Engineers [2.11].

Table 2-4. ASTM Methanol and Ethanol Volatility Classes

Properties	Class 1	Class 2	Class 3
Fuel Methanol (M70-M85)			
Methanol + higher alcohols, min, volume%	84	80	70
Hydrocarbon/aliphatic ether, volume%	14-16	14-20	14-30
Vapor pressure, kPa @ 38°C (psi @ 100°F)	48-62 (7.0-9.0)	62-83 (9.0-12.0)	83-103 (12.0-15.0)
Fuel Ethanol (E_d75-E_d85)			
Ethanol + higher alcohols, min, volume%	79	74	70
Hydrocarbon/aliphatic ether, volume%	17-21	17-26	17-30
Vapor pressure, kPa @ 38°C (psi @ 100°F)	38-59 (5.5-8.5)	48-65 (7.0-9.5)	66-83 (9.5-12.0)

Natural Gas

Natural gas is composed primarily of methane, the smallest and simplest hydrocarbon. One molecule of methane contains one atom of carbon and four atoms of hydrogen, yielding a hydrogen content of 25%, nearly twice that of typical gasoline (see Table 2-5). Natural gas typically has 5 to 10 volume percent higher hydrocarbons, primarily ethane with decreasing amounts of propane and higher

Table 2-5. Methane Properties Compared to Gasoline and Diesel Fuel

Fuel Property	Natural Gas (Methane)[c]	Gasoline[a]	No. 2 Diesel Fuel[a]
Formula	CH_4	C_4 to C_{12}	C_8 to C_{25}
Molecular Weight	16	100-105	200 (approx.)
Composition, Weight %			
Carbon	75	85-88	84-87
Hydrogen	25	12-15	13-16
Oxygen (oxygenated or reformulated gasolines only)	0	0-4	0
Density, kg/L		0.69-0.79	0.81-0.89
(lb/gal)	see Table 2-6	(5.8-6.6)	(6.7-7.4)
Specific Gravity (Relative Density), 15/15°C	see Table 2-6	0.69-0.79	0.81-0.89
Freezing Point, °C	−182	−40	−40 to −1
(°F)	(−296)	(−40)	(−40 to 30)
Boiling Point, °C	−162	27-225	188-343
(°F)	(−260)	(80-437)	(370-650)
Vapor Pressure, kPa @ 38°C	Not	48-103	<1
(psi @ 100°F)	Applicable	(7-15)	(<0.2)
Specific Heat, kJ/(kg-K)	-	2.0	1.8
(Btu/lb-°F)	-	(0.48)	(0.43)
Viscosity, mPa-s @ 20°C	0.01	0.37-0.44	2.6-4.1
(mm/s @ 68°F)	(-)	(0.5-0.6)	(2.8-5.0)
Water Solubility, 21°C (70°F)			
Water in Fuel, Vol %	Negligible	Negligible	Negligible
Electrical Conductivity,			
mhos/cm	-	1×10^{-14}	1×10^{-12}
Latent Heat of Vaporization,			
kJ/kg	510	349	233
(Btu/lb)	(219)	(150)	(100)
Lower Heating Value, 1000 kJ/L		30-33	35-37
(1000 Btu/gal)	see Table 2-6	(109-119)	(126-131)
Flash Point, °C	−188[d]	−43	74
(°F)	(−306)	(−45)	(165)
Autoignition Temperature, °C	540[e]	257	316
(°F)	(1004)	(495)	(600)
Flammability Limits, Vol %			
Lower	5[d]	1.4	1.0
Higher	15	7.6	6.0

Table 2-5. Methane Properties Compared to Gasoline and Diesel Fuel *(cont.)*

Fuel Property	Natural Gas (Methane)[c]	Gasoline[a]	No. 2 Diesel Fuel[a]
Stoichiometric Air-Fuel Ratio, Weight	17.2	14.7	14.7
Flame Spread Rate, m/s (ft/sec)	Not Applicable	4-6[b] (13-20)	- -
Flame Visibility	Visible in all Conditions[b]	Visible in all Conditions[b]	Visible in all Conditions[b]
Octane Number			
Research	120 (Estimated)	88-100	-
Motor	120 (Estimated)	80-90	-
Cetane Number	-	-	40-55

[a] All properties are for conventional, non-oxygenated gasoline and diesel fuel from Ref. [2.17], except where noted. (Numbers include an update in progress as of February, 1997.)
[b] Ref. [2.18]
[c] Ref. [2.21]
[d] Ref. [2.22]
[e] Ref. [2.23]

hydrocarbons (Table 2-5 assumes that natural gas is all methane). When used as a transportation vehicle fuel, natural gas must be either compressed or liquefied to store enough onboard to obtain a practical operating range. The following discusses the generic properties of methane, followed by discussion of typical compressed natural gas (CNG) and liquefied natural gas (LNG).

The boiling point of methane is very low (–162°C [–260°F]) and is not an issue except when dealing with LNG. Water solubility is negligible, though water vapor may be present in natural gas. LNG is inherently water-free since all water is removed during the liquefaction process.

The latent heat of vaporization is a concern only when vaporizing LNG, and that is usually not a concern since obtaining sufficient heat to vaporize LNG is not difficult given the very large temperature difference between LNG and typical ambient temperatures.

The flash point of methane is very low compared to gasoline, though it is hardly a practical consideration since it is very difficult to conceive of situations where pools of liquid methane would exist.

61

The autoignition temperature of methane is about twice as high as that of gasoline, suggesting less chance of ignition due to contact with hot surfaces. However, comparison of autoignition between gaseous and liquid fuels must take into account the latent heat of vaporization of the liquid fuels that can influence the temperature achieved since some energy will be absorbed changing the liquid to a gas. The high autoignition temperature of methane may not be as good an indicator of relative safety as it may first seem.

The flammability limits of methane are higher and wider than those of gasoline or diesel fuel. Here again, comparisons are difficult since methane rises when released in air while gasoline vapors are heavier than air (diesel fuel does not produce significant vapors without being heated). Ignition sources close to the ground that are a concern for gasoline vapors are not a concern for methane that is rising. Given these differences, it takes more methane to create a flammable mixture, and more to exceed the rich flammability limit than gasoline.

The stoichiometric ratio of methane is higher than gasoline or diesel fuel because methane has a higher percentage of hydrogen.

Methane flames are visible under all conditions, just like gasoline and diesel fuel flames, which makes them easy to detect.

Both the research and motor octane rating of methane is something greater than 120, though the exact amount is difficult to measure since the octane tests are devised for liquid fuels. Comparisons can be made with gasoline fuels of high octane, though definitive measurements have not been made to determine accurate values of motor and research octane for methane. The presence of higher hydrocarbons in natural gas can quickly degrade its octane rating relative to methane, and especially with weathering[2] of LNG, which will be discussed in the following sections.

Compressed Natural Gas

To be practical as a transportation fuel, natural gas must be compressed or liquefied to decrease its storage volume. In the U.S., the three common pressures for

---

[2] See Liquefied Natural Gas section.

CNG are: 2400, 3000, and 3600 psi. The lowest of these, 2400 psi, is losing popularity to 3000 and 3600 psi in newer CNG vehicles. Table 2-6 lists the energy storage density of each of these pressures in energy units and in comparison to equivalent gallons of gasoline. For these three CNG pressures, the equivalent amount of gasoline energy stored per unit volume is 22%, 27%, and 33% that of gasoline, respectively. These numbers indicate that CNG takes 4.5, 3.7, and 3.0 times as much volume for storage than gasoline takes. Added to this must be the packing inefficiency of CNG storage cylinders compared to gasoline tanks. CNG cylinders are cylindrical and have much thicker walls than gasoline tanks which lowers the overall amount of fuel that can be stored within a given volume.

Table 2-6. CNG and LNG Energy Storage Densities

Fuel Property	CNG	LNG
Temperature °C (°F)	ambient	−162 (−260)
Energy Content Per Gallon, 1000 Btu/L(1000 Btu/gal)	6.6 (25) @ 16,600 kPa (2400 psi); 22% gasoline equivalent	20 (76); 67% gasoline equivalent
	8.2 (31) @ 20,700 kPa (3000 psi); 27% gasoline equivalent	
	10.0 (38) @ 25,000 kPa (3600 psi); 33% gasoline equivalent	
Gasoline Gallon Equivalence	3500 liters (125 scf)	5.7 liters (1.5 gallons)

A very large difference between CNG cylinders and gasoline tanks is that many gasoline tanks are made from plastics that can be molded to fit very irregular shapes within the vehicle structure while CNG cylinders cannot. For these reasons, the practical energy storage density of CNG within a vehicle is lower than the comparison of energy storage density indicates.

Fuels stored at high pressures present some unique handling considerations. For CNG, two are notable: Joule-Thompson cooling and static electricity build-up. The Joule-Thompson effect is important because in CNG fuel systems, natural gas often undergoes rapid decrease in pressure as it moves through the fuel system. The decreases in pressure cause localized decreases in temperature that can approach the boiling point of methane (−162°C [−260°F]). The cooling effect is most pronounced in the pressure drop through fittings and pressure regulators. Water vapor present will obviously freeze in such situations, and the presence of sulfur and compressor oil carryover also can cause problems. Water vapor and sulfur can combine to form hydrates in CNG fuel systems where Joule-Thompson cooling occurs. Hydrates are crystalline in structure, similar to snow, and can easily obstruct CNG fuel lines. Prevention of hydrates requires limitation of water vapor and sulfur contents, or localized heating of components. Oil carried over from CNG compressors can cause problems because at low temperatures, it too can become very viscous with potential to clog lines or cause components such as pressure regulators to malfunction.

Liquefied Natural Gas

The advantage of LNG in terms of energy storage density is readily evident from Table 2-6 and is the reason over-the-road truckers prefer LNG to CNG. (LNG fuel systems also are lighter per unit volume of fuel storage compared with CNG fuel systems.) LNG is a very clean fuel since no water vapor or sulfur compounds can survive the liquefaction process. (This is why LNG is not odorized.) Higher hydrocarbons such as ethane and propane can be present, though this is usually undesirable because of a phenomenon called "weathering." LNG is stored at very low temperatures (−120 to −162°C [−195 to −260°F]). While the insulation of LNG storage tanks is very good, LNG still experiences a net gain of heat because of the difference between the fuel temperature and the ambient temperature. As heat is gained by the LNG, vapor is generated which must eventually be released from the storage tank to avoid over-pressurization because LNG tanks are not designed to maintain high pressures (if they were, they would lose the advantages in tank weight which come from storing natural gas as a liquid rather than a gas). The methane content of the vapor released from the fuel is higher than that of the fuel itself because most of the other compounds contained in natural gas have higher boiling points than methane, and methane typically makes up 95% or greater of the composition of natural gas.[3] The loss of high-methane content vapor during storage causes the methane content of the LNG to decline over time. The loss of methane vapor and the deterioration of methane content in the LNG is referred to as "LNG weathering" [2.12].

---

[3] The non-methane components of natural gas are predominantly ethane and propane, though nitrogen and carbon dioxide may also be present. Of these, nitrogen is the only one with a boiling point lower than methane.

The energy density of LNG is about 67% that of gasoline (Table 2-6) and 59% that of diesel fuel. This suggests that it takes 1.5 liters of LNG to equal a liter of gasoline and 1.7 liters of LNG to equal a liter of diesel fuel. Like CNG, LNG tanks have a smaller fuel storage volume compared to total exterior volume than the typical diesel fuel tank because of the insulation required to keep the LNG cold.

LP Gas

LP gas is a generic term for fuels that include butane, propane, and small amounts of other hydrocarbons. The common characteristic among these fuels is that they are easily liquefied by the application of modest pressures (less than 300 psi). LP gas is widely used in rural areas of the country as a home cooking and heating fuel, and for use in outdoor home barbecues. For fuel use, the industry has standardized on a specification that requires 90% propane minimum, 2.5% butane maximum, and 5% propylene maximum, known as HD-5. Since this discussion centers on use of LP gas as a transportation fuel—which means LP gas conforming to the HD-5 specification—propane will be used instead of LP gas.

The molecular weight of propane is nearly half that of the typical gasoline molecule (see Table 2-7). The characteristics of propane lie between those of methane and gasoline, with propane having slightly more hydrogen and less carbon than gasoline. As a liquid, propane weighs about 68% that of gasoline. As might be expected, propane has a freezing point that is well below any typical ambient temperature. Its boiling point is also very low, which is typical of lower-molecular-weight hydrocarbons. Its vapor pressure at 37.8°C (100°F) is 1303 kPa (189 psi) which is much beyond that of gasoline. Propane will absorb a small amount of water, though whether this is more or less than gasoline depends on the aromatic content of the gasoline.

Propane has about a 20% increase in latent heat of vaporization compared with gasoline, though the practical significance of this difference is debatable. The heating value (lower) of propane is 72% that of typical gasoline. This is more than the increase in specific weight would suggest and is caused by propane's higher energy content per pound than gasoline. Even so, it takes nearly 1.4 gallons of propane to equal 1.0 gallon of gasoline.

Table 2-7. Propane Properties Compared to Gasoline and Diesel Fuel

Fuel Property	Propane[c]	Gasoline[a]	No. 2 Diesel Fuel[a]
Formula	C_3H_8	C_4 to C_{12}	C_8 to C_{25}
Molecular Weight	44.09	100-105	200 (approx.)
Composition, Weight %			
Carbon	82	85-88	84-87
Hydrogen	18	12-15	13-16
Oxygen (oxygenated or reformulated gasolines only)	0	0-4	0
Density, kg/L, 15/15°C	0.50	0.69-0.79	0.81-0.89
(lb/gal), 60/60°F	4.2	(5.8-6.6)	(6.7-7.4)
Specific Gravity (Relative Density), 15/15°C	0.5	0.69-0.79	0.81-0.89
Freezing Point, °C	−187	−40	−40 to −1
(°F)	(−306)	(−40)	(−40 to 30)
Boiling Point, °C	−42	27-225	188-343
(°F)	(−44)	(80-437)	(370-650)
Vapor Pressure, kPa @ 38°C	1303	48-103	<1
(psi @ 100°F)	(189)	(7-15)	(<0.2)
Specific Heat, kJ/(kg-K)	2.48	2.0	1.8
(Btu/lb-°F)	(0.592)	(0.48)	(0.43)
Viscosity, mPa-s @ 20°C	0.102	0.37-0.44	2.6-4.1
(mm/s @ 68°F)		(0.5-0.6)	(2.8-5.0)
Water Solubility, 21°C (70°F)			
Water in Fuel, Vol %	0.065[e]	Negligible	Negligible
Electrical Conductivity			
mhos/cm	-	1×10^{-14}	1×10^{-12}
Latent Heat of Vaporization,			
kJ/kg	426	349	233
(Btu/lb)	(183)	(150)	(100)
Lower Heating Value, 1000 kJ/L	23	30-33	35-37
(1000 Btu/gal)	(82.4)	(109-119)	(126-131)
Flash Point, °C	−104[d]	−43	74
(°F)	(−156)	(−45)	(165)
Autoignition Temperature, °C	457[f]	257	316
(°F)	(855)	(495)	(600)
Flammability Limits, Vol %			
Lower	2.1[f]	1.4	1.0
Higher	9.5	7.6	6.0

**Table 2-7. Propane Properties Compared to Gasoline
and Diesel Fuel *(cont.)***

Fuel Property	Propane[c]	Gasoline[a]	No. 2 Diesel Fuel[a]
Stoichiometric Air-Fuel Ratio, Weight	15.7	14.7	14.7
Flame Spread Rate, m/s (ft/sec)	Not Applicable	4-6[b] (13-20)	- -
Flame Visibility	Visible in all Conditions[b]	Visible in all Conditions[b]	Visible in all Conditions[b]
Octane Number Research Motor	112 97	88-100 80-90	- -
Cetane Number	-	-	40-55

[a] All properties are for conventional, non-oxygenated gasoline and diesel fuel from Ref. [2.17], except where noted. (Numbers include an update in progress as of February, 1997.)
[b] Ref. [2.18]
[c] All properties from Ref. [2.24].
[d] Ref. [2.25]
[e] Ref. [2.22]
[f] Ref. [2.23]

The flash point of propane is very low, in line with its very low freezing and boiling points. Because propane's flash point is very low, its practical significance is reduced since propane will almost always be at temperatures high above its flash point. The autoignition temperature of propane is nearly twice as high as for gasoline, giving it some additional margin of safety for igniting when exposed to hot surfaces. The flammability limits of propane are higher than for gasoline, but not significantly so. In practical use, the difference in flammability limits for propane are not significantly different from those for gasoline. The stoichiometric air-fuel ratio for propane is just slightly higher than for gasoline, reflecting propane's higher hydrogen content. Like gasoline, propane flames are visible under all conditions, making flame visibility a non-issue.

The octane ratings of propane are significantly higher than for gasoline, and engines designed for propane will usually have higher compression ratios to take advantage of its higher octane.

Vegetable Oils

Vegetable oils can be used as diesel engine fuels but they have high viscosity and high temperature pour points relative to diesel fuel. The high viscosity of vegetable oils tends to alter the injector spray pattern inside the engine causing fuel impingement on the piston and other combustion chamber surfaces. This leads to formation of carbon deposits in the engine that eventually result in problems such as stuck piston rings with subsequent engine failures that wouldn't occur when using diesel fuel [2.13]. These characteristics make vegetable oils impractical for use as transportation fuels in most climates and in most engines.

The undesirable characteristics of vegetable oils can be substantially changed by replacing triglyceride molecules in them with lighter alcohol molecules such as methanol or ethanol. This reaction is carried out in the presence of a catalyst and produces glycerol in addition to transesterfied vegetable oils that are given the generic name of "biodiesel." Most vegetable oils and several common animal fats can be modified in this manner which greatly improves their properties as fuels. The vegetable oils most often considered for transesterification include soybean oil, rapeseed oil, and sunflower oil. Table 2-8 lists the properties of soybean and rapeseed methyl esters (the names given to the modified versions of these vegetable oils) compared to gasoline and diesel fuel. (Many properties for methyl esters are not yet available since methyl esters have only recently been considered for use as alternative transportation fuels.)

Soybean and rapeseed methyl esters both are made up of a fairly narrow range of hydrocarbon molecules containing mostly 18 or 19 carbons, with molecular weights around 300. In comparison, gasoline contains mostly lighter hydrocarbons, and while diesel fuel contains heavier molecules, it has a much wider range of hydrocarbons. The most significant difference in the composition of the methyl esters compared to gasoline and diesel fuel is that they contain oxygen from the carboxyl group present in their structure which was created by the reaction with methanol. The oxygen content of the methyl esters is usually around 10% or 11%, but the rapeseed methyl ester in Table 2-8 has been modified by removing some of the higher pour point esters to improve its handling properties at low temperatures which resulted in lowering the percentage of oxygen present in the fuel [2.14].

Table 2-8. Vegetable Oil Properties Compared to Gasoline and Diesel Fuel

Fuel Property	Soybean Methyl Ester[d]	Rapeseed Methyl Ester[c]	Gasoline[a]	No. 2 Diesel Fuel[a]
Formula	C_{18} to C_{19}	C_{18} to C_{19}	C_4 to C_{12}	C_8 to C_{25}
Molecular Weight	300 (approx.)	300 (approx.)	100-105	200 (approx.)
Composition, Weight %				
Carbon	78	81	85-88	84-87
Hydrogen	11	12	12-15	13-16
Oxygen	11	7	0-4	0
Density, kg/L, 15/15°C	0.87	0.88	0.69-0.79	0.81-0.89
(lb/gal), 60/60°F	(7.3)	(7.3)	(5.8-6.6)	(6.7-7.4)
Specific Gravity	0.87	0.88	0.69-0.79	0.81-0.89
Pour Point, °C	–3	–15	-	–23[c]
(°F)	(27)	(5)	-	(–10)
Boiling Point, °C	-	~350	27-225	188-343
(°F)	-	(~662)	(80-437)	(370-650)
Vapor Pressure, kPa @ 38°C	<1	<1	48-103	<1
(psi @ 100°F)	(<0.2)	(<0.2)	(7-15)	(<0.2)
Specific Heat, kJ/(kg-K)	-	-	2.0	1.8
(Btu/lb-°F)	-	-	(0.48)	(0.43)
Viscosity, mPa-s @ 20°C	3-6	3-6	0.37-0.44	2.6-4.1
Water Solubility, 21°C (70°F)				
Water in Fuel, ppm	-	850	<50	<50
Electrical Conductivity, mhos/cm	-	-	1×10^{-14}	1×10^{-12}
Latent Heat of Vaporization, kJ/kg	-	-	349	233
(Btu/lb)	-	-	(150)	(100)
Lower Heating Value, 1000 kJ/L	~32	~37	30-33	35-37
(1000 Btu/gal)	(~117)	(~135)	(109-119)	(126-131)
Flash Point, °C	-	179	–43	74
(°F)	-	(354)	(–45)	(165)
Autoignition Temperature, °C	-	-	257	316
(°F)	-	-	(495)	(600)
Flammability Limits, Vol %				
Lower	-	-	1.4	1.0
Higher	-	-	7.6	6.0
Stoichiometric Air-Fuel Ratio, Weight	-	-	14.7	14.7
Flame Spread Rate, m/s	-	-	4-6[b]	-
(ft/sec)	-	-	13-20	-

Table 2-8. Vegetable Oil Properties Compared to Gasoline and Diesel Fuel *(cont.)*

Fuel Property	Soybean Methyl Ester[d]	Rapeseed Methyl Ester[c]	Gasoline[a]	No. 2 Diesel Fuel[a]
Flame Visibility	Visible in all Conditions	Visible in all Conditions	Visible in all Conditions[b]	Visible in all Conditions[b]
Octane Number				
Research	-	-	88-100	-
Motor	-	-	80-90	-
Cetane Number	52	62	-	40-55

[a] All properties are for conventional, non-oxygenated gasoline and diesel fuel from Ref. [2.17], except where noted. (Numbers include an update in progress as of February, 1997.)
[b] Ref. [2.18]
[c] Rapeseed Methyl Ester properties from Ref. [2.14].
[d] Soybean Methyl Ester properties from Ref. [2.26].

Soybean and rapeseed methyl esters both have densities similar to diesel fuel. Their pour points are not as favorable. Straight soybean methyl ester has a very high pour point (–3°C [–27°F]) which would cause problems for vehicles in most non-tropical climates. The rapeseed methyl ester of Table 2-8 illustrates the improvement in pour point possible just by removing some of the esters that have higher pour points. Additives would no doubt further improve the pour point characteristics of these fuels.

The boiling point of methyl esters is around 350°C (662°F), which is similar to the highest-boiling-point hydrocarbons contained in diesel fuel. The methyl esters have a very narrow boiling range and act similarly to single constituent fuels such as methanol. With this narrow boiling range at high temperature, the low vapor pressure of methyl esters is hardly a surprise. The flash point of methyl esters is much higher than that of diesel fuel. Combined together, the high boiling point and high flash point make methyl esters as safe or safer than diesel fuels from ignition due to fuel spills. The methyl esters have visible flames, like diesel fuel, so lack of flame visibility is not a safety concern.

The viscosity of methyl esters is similar to diesel fuel, and in most cases higher than diesel fuel. This could potentially have an impact on the quantity of fuel injected by diesel engine fuel injection systems, but it is not likely to cause significant changes.

Methyl esters have a much greater water solubility than diesel fuel, no doubt due to the polar nature of the methyl groups they contain. The soybean methyl ester of Table 2-8 has a heating value similar to gasoline while the rapeseed methyl ester has a heating value similar to diesel fuel. The difference is most likely due to the larger oxygen content of the soybean methyl ester compared to the rapeseed methyl ester.

Methyl esters have good cetane numbers, spanning a range that starts with the high end of the best diesel fuels to values in the low 60s. High cetane numbers are conducive to low engine operating noise and good starting characteristics, though the latter is offset with methyl esters due to their high viscosity and high pour points.

The American Society for Testing and Materials (ASTM) initiated a Biodiesel Task Force in June 1994 to develop a specification for vegetable ester fuels for use in diesel-engine-powered vehicles. When completed, this specification for vegetable ester fuels will be the analog of ASTM D 975, the specification for petroleum-based diesel fuel.

Hydrogen

Hydrogen is the only alternative fuel that does not contain any carbon or oxygen. It is the lightest fuel possible, with a molecular weight of only 2.02. Even as a liquid, hydrogen is only about one-tenth the weight per liter of gasoline (but has about one-quarter the energy).

Liquefied hydrogen (LH_2) is a cryogenic liquid—its boiling point is –253°C (–423°F). Thus, the storage containers for LH_2 must have the best insulation available. The cold temperature of LH_2 requires storage tanks made from stainless steel. The insulation technology required to keep LH_2 from boiling, combined with the requirement to use stainless steel, make storage containers for LH_2 expensive compared to liquid fuels such as gasoline, diesel fuel, methanol, or ethanol. Storage containers for LH_2 are even more expensive than for LNG because the insulation requirements are more severe.

The specific heat of hydrogen is much higher than that of gasoline or diesel fuel, but because hydrogen is so light, the net effect is essentially negligible. The latent heat of vaporization of hydrogen is 28% higher than gasoline and 92% higher than diesel fuel, though the practical significance of this difference is debatable. Hydrogen vehicle fuel systems will likely rely on gaseous hydrogen, even if it is stored onboard as a liquid. The heat absorbed to vaporize each 1000 kJ of hydrogen is only about 40% that of gasoline. So, even though hydrogen has a higher latent heat of vaporization, its effect overall is less than that of gasoline.

Hydrogen as a liquid has about 27% of the energy per liter of gasoline, and about 23% of the energy per liter of diesel fuel. As a compressed gas at 3000 psi, hydrogen has only about 5% the energy of gasoline per liter. Thus, to equal the energy storage of gasoline, hydrogen will need at least 4 times the fuel storage volume if stored as a liquid, and 20 times the storage volume if stored as a compressed gas.

Hydrogen has the widest flammability range of all fuels—from 4 to 75 volume percent in air. This wide flammability range has significant implications for hydrogen safety. Hydrogen also tends to diffuse more readily than natural gas, so leaks of hydrogen will tend to diffuse rapidly within a space and will be in the flammable range for a long time in comparison to other fuels [2.15]. Hydrogen also burns without a visible flame in direct sunlight, which is an additional safety concern.

Though the octane and cetane values of hydrogen have not been well defined, much work has been done on spark-ignition internal-combustion engines to use hydrogen as a fuel [2.16]. It has been discovered that spark-ignition internal-combustion engines using hydrogen are prone to a phenomenon called "flashback," where backfires through the intake system occur randomly with great force due to hydrogen's fast flame speed. Flashback is believed to be caused by preignition, and the only sure solution known at present is to use direct cylinder injection of hydrogen. For these reasons, octane value of hydrogen is not as important as designing an engine that will avoid flashback. The engine tests performed to date have used compression ratios typical of those used for gasoline engines, so hydrogen engines should not have a disadvantage in terms of basic engine thermal efficiency. Hydrogen can also be combusted very lean, which gives it an efficiency advantage over gasoline engines that must rely on stoichiometric mixtures and catalytic control of the exhaust gases for emissions control.

Table 2-9. Hydrogen Properties Compared to Gasoline and Diesel Fuel

Fuel Property	Hydrogen[c]	Gasoline[a]	No. 2 Diesel Fuel[a]
Formula	H_2	C_4 to C_{12}	C_8 to C_{25}
Molecular Weight	2.02	100-105	200 (approx.)
Composition, Weight %			
Carbon	0	85-88	84-87
Hydrogen	100	12-15	13-16
Oxygen (oxygenated or reformulated gasolines only)	0	0-4	0
Density, kg/L, 15/15°C (g=gas; l=liquid)	0.0013(g), 0.07(l)	0.69-0.79	0.81-0.89
(lb/gal), 60/60°F	(0.010 g) (0.6 l)	(5.8-6.6)	(6.7-7.4)
Specific Gravity	0.07(g), 0.07(l)	0.69-0.79	0.81-0.89
Freezing Point, °C	–275	–40	–40 to –1
(°F)	(–463)	(–40)	(–40 to 30)
Boiling Point, °C	–253	27-225	188-343
(°F)	(–423)	(80-437)	(370-650)
Vapor Pressure, kPa @ 38°C (psi @ 100°F)	Not Applicable	48-103 (7-15)	<1 (<0.2)
Specific Heat, kJ/(kg-K)	14.2	2.0	1.8
(Btu/lb-°F)	(3.39)	(0.48)	(0.43)
Viscosity, mPa-s @ 20°C	0.009	0.37-0.44	2.6-4.1
Water Solubility, 21°C (70°F)			
Water in Fuel, Vol %	Negligible	Negligible	Negligible
Electrical Conductivity, mhos/cm	-	1×10^{-14}	1×10^{-12}
Latent Heat of Vaporization, kJ/kg	448	349	233
(Btu/lb)	(192.7)[g]	(150)	(100)
Lower Heating Value, 1000 kJ/L (liq.)	8.4	30-33	35-37
(1000 Btu/gal) (liq.)	(30)	(109-119)	(126-131)
Flash Point, °C	-	–43	74
(°F)	-	(–45)	(165)
Autoignition Temperature, °C	-	257	316
(°F)	-	(495)	(600)
Flammability Limits, Vol %			
Lower	4[g]	1.4	1.0
Higher	75	7.6	6.0

**Table 2-9. Hydrogen Properties Compared to Gasoline
and Diesel Fuel *(cont.)***

Fuel Property	Hydrogen[c]	Gasoline[a]	No. 2 Diesel Fuel[a]
Stoichiometric Air-Fuel Ratio, Weight	34.3	14.7	14.7
Flame Spread Rate, m/s (ft/sec)	-	4-6[b]	-
	-	13-20	-
Flame Visibility	Invisible in direct sunlight	Visible in all Conditions[b]	Visible in all Conditions[b]
Octane Number			
Research	-	88-100	-
Motor	-	80-90	-
Cetane Number	-	-	40-55

[a] All properties are for conventional, non-oxygenated gasoline and diesel fuel from Ref. [2.17], except where noted. (Numbers include an update in progress as of February, 1997.)
[b] Ref. [2.18]
[c] All properties from Ref. [2.27], except where noted.
[d] Ref. [2.25]
[e] Ref. [2.22]
[f] Ref. [2.23]
[g] Ref. [2.28]

Hydrogen is the preferred fuel for fuel cell vehicles. Fuel cells have higher peak efficiency than current spark-ignition or diesel engines, and have essentially no emissions other than water vapor. However, the low energy storage density of hydrogen limits fuel cell vehicle operating range, and hydrogen is currently more expensive than most conventional or alternative fuels. For these reasons, advocates of fuel cell vehicles are attempting to reform methanol or gasoline onboard to produce hydrogen for fuel cells. This lowers vehicle fuel efficiency and causes emissions but could be a bridge to hydrogen vehicles of the future.

Sources of Additional Information

The following are sources of additional information about transportation fuels:

- ASTM D 4814, "Standard Specification for Automotive Spark-Ignition Fuel, American Society for Testing and Materials," 100 Barr Harbor Drive, West Conshohocken, Pa. 19428-2959, 610-832-9585.

- ASTM D 975, "Specification for Diesel Fuel Oils," American Society for Testing and Materials, 100 Barr Harbor Drive, West Conshohocken, Pa. 19428-2959, 610-832-9585.

- ASTM D 1835, "Specification for Liquefied Petroleum (LP) Gases," American Society for Testing and Materials, 100 Barr Harbor Drive, West Conshohocken, Pa. 19428-2959, 610-832-9585.

- ASTM D 5797, "Specification for Fuel Methanol M70-M85 for Automotive Spark-Ignition Engines," American Society for Testing and Materials, 100 Barr Harbor Drive, West Conshohocken, Pa. 19428-2959, 610-832-9585.

- ASTM D 5798, "Specification for Fuel Ethanol E_d70-E_d85 for Automotive Spark-Ignition Engines," American Society for Testing and Materials, 100 Barr Harbor Drive, West Conshohocken, Pa. 19428-2959, 610-832-9585.

- SAE J1616, "Recommended Practice for Compressed Natural Gas Fuel," Society of Automotive Engineers, 400 Commonwealth Drive, Warrendale, Pa. 15096-0001, 412-776-4841.

- Brinkman, N.D., Halsall, R., Jorgensen, S.W., and Kirwan, J.E., "The Development of Improved Fuel Specifications for Methanol (M85) and Ethanol (E_d85)," SAE Paper No. 940764, Society of Automotive Engineers, 400 Commonwealth Drive, Warrendale, Pa. 15096-0001, 412-776-4841.

References

2.1. National Fire Protection Association, *Fire Hazard Properties of Flammable Liquids, Gases, and Volatile Solids*, NFPA 325M-1984, 1984 Ed.

2.2. Heywood, John B., *Internal-Combustion Engine Fundamentals*, McGraw-Hill Book Company, 1988.

2.3. *Dictionary of Scientific and Technical Terms*, Third Ed., McGraw-Hill Book Company, 1984.

2.4. Owen, K., and Coley, T., *Automotive Fuels Handbook*, R-105, Society of Automotive Engineers, Warrendale, Pa., 1990.

2.5.   ASTM D 323, "Test Method for Vapor Pressure of Petroleum Products," American Society for Testing and Materials, West Conshohocken, Pa.

2.6.   Archer Daniels Midland Co., *Fuel Ethanol—Technical Bulletin*, Decatur, Ill., September 1993.

2.7.   Gülder, Ö.L., Glavincevski, B., and Battista, V., *Visibility of Methanol Pool Flames*, 1993 Windsor Workshop on Alternative Fuels.

2.8.   Hunwartzen, I., "Modification of CFR Test Engine Unit to Determine Octane Numbers of Pure Alcohols and Gasoline-Alcohol Blends," SAE Paper No. 820002, Society of Automotive Engineers, Warrendale, Pa., 1982.

2.9.   ASTM D 5797, "Specification for Fuel Methanol M70-M85 for Automotive Spark-Ignition Engines," American Society for Testing and Materials, West Conshohocken, Pa.

2.10. ASTM D 5798, "Specification for Fuel Ethanol E_d70-E_d85 for Automotive Spark-Ignition Engines," American Society for Testing and Materials, West Conshohocken, Pa.

2.11. SAE Cooperative Research Program Project Group 3, "A Discussion of M85 (85% Methanol) Fuel Specifications and Their Significance," Report No. CRP-2, Society of Automotive Engineers, Warrendale, Pa., September 1991.

2.12. Gibbs, J.L., Bechtold, R.L., and Collison, C.E., "The Effects of LNG Weathering on Fuel Composition and Vehicle Management Techniques," SAE Paper No. 952607, Society of Automotive Engineers, Warrendale, Pa., 1995.

2.13. McDonald, J., Purcell, D.L., McClure, B.T., and Kittelson, D.B., "Emissions Characteristics of Soy Methyl Ester Fuels in an IDI Compression Ignition Engine," SAE Paper No. 950400, Society of Automotive Engineers, Warrendale, Pa., 1995.

2.14. Reece, D.L., and Peterson, C.L., "Biodiesel Testing in Two On-Road Pickups," SAE Paper No. 952757, Society of Automotive Engineers, Warrendale, Pa., 1995.

2.15. Swain, M.R., and Swain, M.N., "Passive Ventilation Systems for the Safe Use of Hydrogen," submitted for publication in *International Journal of Hydrogen Energy*, February 1996.

2.16. Swain, M.R., Adt, R.R., Jr., and Pappas, J.M., "Experimental Hydrogen-Fueled Automotive Engine Design Data-Base Project," U.S. Department of Energy Report DOE/CS/51212-1, Vol. 3, May 1983.

2.17. American Petroleum Institute, *Alcohols and Ethers*, API Publication 4261, July 1988.

2.18. Machiele, Paul A., "Flammability and Toxicity Tradeoffs with Methanol Fuels," SAE Paper No. 872064, Society of Automotive Engineers, Warrendale, Pa., 1987.

2.19. Mueller Associates, *Methanol Storage and Handling Factsheet*, May 1987.

2.20. Kopp, V.R., *et al.*, "Test Fuel Blending and Analysis for Phase II Follow-Up Programs: The Auto/Oil Air Quality Improvement Program," SAE Paper No. 952506, Society of Automotive Engineers, Warrendale, Pa., 1995.

2.21. Obert, E.F., *Internal-Combustion Engines and Air Pollution*, Third Ed., Intext Educational Publishers, New York, N.Y.

2.22. Compressed Gas Association, *Handbook of Compressed Gas*, Second Ed., 1981.

2.23. Los Alamos National Laboratory, *Gaseous Fuel Safety Assessment for Light-Duty Automotive Vehicles*, LA-9829-MS, November 1983.

2.24. *Handbook of Butane-Propane Gases*, Butane-Propane News, Arcadia, Ca., 1973.

2.25. The Aerospace Corporation, *Assessment of Methane-Related Fuels for Automotive Fleet Vehicles*, U.S. Department of Energy, DOE/CE/50179-1, February 1982.

2.26. Zhang, Y. and Van Gerpen, J.H., "Combustion Analysis of Esters of Soybean Oil in a Diesel Engine," SAE Paper No. 960765, Society of Automotive Engineers, Warrendale, Pa., 1996.

2.27. *Handbook of Chemistry and Physics*, 44th Ed., Chemical Rubber Publishing Co., 1962.

2.28. National Fire Protection Association, *Fire Protection Handbook*, 13th Ed., 1969.

Chapter Three

Materials Compatibility

Our highway transportation vehicles have evolved over the years to be very reliable and much of the credit for that is due to the refinements made in storing and handling fuels. Compatibility of the materials used in engine fuel systems and bulk transport, storage, and dispensing systems has always been a concern and remains important as these systems and fuels evolve over time. The primary areas of concern for materials compatibility include corrosion of metals and impacts on the elastomers[1] used in the engine, primarily in the fuel system. The elastomer impacts include swelling, shrinking, hardening, cracking, and other changes that lead to failure of the part using the elastomers. Elastomers used in fuel lines may fail, causing fuel leaks, or become porous and contribute to evaporative emissions. A growing concern for materials compatibility is the increasing amount of plastics used in vehicle fuel systems. Many fuel tanks have switched to plastic to allow more efficient use of the very limited and irregular volumes that are available in vehicles to place fuel tanks. Plastic fuel tanks that become brittle or porous over time due to interactions with fuel are not desirable.

While fuels are usually the source of materials compatibility problems, they also can provide some of the solutions through the use of additives. Additives from several manufacturers are available to help prevent corrosion of fuel system components and to stabilize fuel properties from oxidation. While fuel additives are helpful, they can rarely do it all themselves and must be developed in concert with knowledge of fuel system materials properties.

---

[1] Elastomer is a term used to describe a wide range of "soft" materials used in fuel systems. One of the best examples is the common O-ring—usually made from a soft, rubber-like material—that depends on its ability to deform under pressure to perform properly. The most common job for elastomers in fuel systems is to prevent leaks from fittings and joints.

The following sections address materials compatibility for alternative fuels. In many cases, alternative fuels bring unique materials compatibility problems. Users of alternative fuel vehicles should be aware of these problems to help prevent mistakes in setting up refueling systems and to recognize and identify materials compatibility problems as they occur.

The Alcohols

Methanol and ethanol are the only alcohols considered for use as fuels where they are the major components. Higher alcohols and ethers (such as isopropyl alcohol, tertiary butyl alcohol, methyl tertiary butyl ether, ethyl tertiary butyl ether, di-isopropyl ether, and others) are being used as blending components in gasoline, but they pose little if any change in materials compatibility requirements from those of the base gasoline itself. The following sections present materials compatibility issues for methanol and ethanol in neat form and when 15 volume percent gasoline is added to make M85 and E85.

Methanol

Of the alcohols in pure form, methanol is the most aggressive in terms of materials compatibility, affecting both metals and elastomers. Of the metals, magnesium has the worst reaction when in contact with methanol. Methanol will very rapidly corrode magnesium; the use of magnesium is not recommended in applications where contact with methanol is possible. Aluminum is also corroded by contact with methanol, but the reaction is slower than with magnesium and occurs more rapidly to aluminum exposed to methanol vapor than methanol liquid. The corrosion products of aluminum and methanol (aluminum hydroxide) are gelatinous precipitates and will plug filters and cause operating problems with fuel injectors and increased engine wear [3.1]. General Motors has determined that aluminum hydroxide concentrations of 0.1 mg/L are the limit to ensure vehicle fuel filter life of 40,000 km (25,000 miles) [3.2].

Transport of methanol in aluminum truck tankers demonstrates the slow reaction of aluminum to methanol. Analysis of M85 transported in aluminum tankers has proven that the amount of corrosion that takes place during the few days of transport is insignificant [3.3]. The fact that the interior of the tank was constantly wetted no doubt contributed to this result. If aluminum tankers are allowed to sit idle following a methanol delivery, corrosion may accelerate and become unacceptable. Further experience is needed to verify whether transport in aluminum tankers can be considered an acceptable practice.

Ordinary carbon steel is surprisingly one of the metals least affected by methanol that is "dry" (does not contain significant amounts of dissolved water) and free from foreign matter and impurities. (Galvanized steel should not be used, as methanol will strip the galvanizing which will then contaminate the fuel.) The methanol industry employs bulk methanol storage tanks made from carbon steel, and most of the railcars that transport methanol are made from carbon steel. Methanol containing significant amounts of water and impurities (especially if they contribute chloride or sulfate ions to the methanol and water solution) will cause carbon steel to rust. Contamination of methanol with chloride ions is common since chlorinated solvents are often distributed using the same trucks and containers. Other impurities that cause methanol to become more corrosive include sodium formate, organic peroxides, sulfur compounds, and reactive metals [3.4].

Stainless steel is the metal least affected by methanol. Most of the experience to date has been with the higher grade stainless steel alloys such as 304. It is unknown how well alloys such as 409 might work, but given the good performance of carbon steels with methanol, it is likely that components made from 409 would fare well.

Brass, bronze, and die cast zinc have been found to be corroded more quickly by methanol than by gasoline [3.5]. Copper is to be avoided, not only because it is likely to be corroded by methanol, but because it is not suitable for hydrocarbon fuels. (Flexible fuel vehicles use both methanol and gasoline, and practical methanol fuels now are anticipated to be part gasoline.)

Several corrosion inhibitors such as polyamide, polyamine, dithiocarbamate, thiophosphate ester, organic acid, sulfide, and selenide types have all been tried in methanol without significant success [3.5]. Similarly, nonmetallic coatings of metals to effectively prevent corrosion have not been developed to date. One application where coatings have proven effective is in prevention of anodic dissociation of fuel pumps immersed in methanol. Since methanol is many times more conductive than gasoline, components such as fuel pumps and electrical fuel level gauges can cause induced currents that in turn remove metal from these components. Polyolefin coatings have been found to be effective in preventing such metal removal [3.5].

Several metallic coatings have been tested to reduce the corrosivity of methanol. Nickel is the only one that has been found to be reliably effective, and can be applied to steel and aluminum through several processes.

Grades of fiberglass that are resistant to methanol have been developed, primarily for use in bulk storage tanks. Fiberglass not designed for methanol will quickly soften and delaminate. Ford has reported that onboard fuel lines made of poly phenylene sulfide resin give good performance [3.6].

Fuel hoses such as those used to dispense methanol into vehicles have been developed that are methanol compatible. Cross-linked polyethylene has been found to be a compatible methanol fuel material. Fuel hoses designed for gasoline will harden and crack while depositing significant amounts of plasticizer into the methanol.

Its highly polar nature makes methanol a very good solvent, which in turn causes materials compatibility problems with elastomers such as swelling, shrinking, hardening, softening, and cracking. High-flourine-content elastomers and teflon have proved to be compatible with methanol. Buna-N, Viton[2], rubber, nitrile, acrylate, polyurethane, and most plastics are attacked by methanol [3.7]. Methanol may also attack cork and methyl methacrylate, and soften fiber used in gaskets [3.5]. Threaded pipe connections are best sealed by using teflon tape, since methanol dissolves most other pipe dopes developed for gasoline and diesel fuel.

Methanol has been found to react with certain lubricating oil additives and cause them to be leached out of the oil. Methanol-compatible lubricating oil packages have been developed and should be used when methanol is the fuel [3.8].

Filter manufacturers have been successful in developing both filter elements and adhesives that are compatible with methanol [3.9].

---

[2] Viton is a trade name for a class of elastomers that will have differing responses to methanol and other fuels.

Ethanol

Ethanol is widely acknowledged to be less aggressive toward metals and elastomers than methanol, but little research and development has been devoted to the specific problems posed by ethanol. Ethanol typically has more water in it than methanol (an artifact of production) which may affect solubility of contaminants and corrosion potential. One ethanol contaminant that can arise from production is acetic acid, which is water-soluble and will corrode some automotive fuel system components. For instance, General Motors found that E85 caused more corrosion in fuel pumps than M85, presumably because of a higher level of dissolved contaminants [3.2]. Since much more development has been devoted to compatibility with methanol fuels, the general approach for ethanol has been to use materials developed for methanol, even though they may be "over-engineered."

The metals recommended for use with ethanol include carbon steel, stainless steel, and bronze [3.10]. Like methanol, metals such as magnesium, zinc castings, brass, and copper are not recommended. Aluminum can be used if the ethanol is very pure, otherwise it should be nickel-plated or suitably protected from corrosion by another means. The metals compatible with ethanol represent a much wider range than those for methanol and represent most of the metals currently used in fuel systems, so few changes would be anticipated when using ethanol.

The previous paragraph assumes that the ethanol will be "dry" (containing no water) and contain only very small amounts of contaminants such as chloride and sulfate ions that would greatly increase the corrosivity of ethanol. Ethanol produced for fuel purposes in the past has contained up to 5 volume percent water and ion concentrations that made it much more corrosive than pure ethanol [3.7]. For an ethanol fuel with these corrosion characteristics, it was found that aluminum and steel could be coated with cadmium, hard chromium, nickel, or anodized aluminum to make them compatible. Coatings such as zinc, lead, and phosphate were found to be inadequate to prevent corrosion [3.7].

Ethanol is much less aggressive to elastomers than methanol. Most common elastomers will work satisfactorily with ethanol including Buna-N, Viton, fluorosilicones, neoprene, and natural rubber [3.10]. Teflon and nylon are compatible and work well with ethanol. Cork will deteriorate badly in contact with ethanol, polyamide plastic will be hardened and darkened, and fiberglass-reinforced plastic laminate will soften and yellow appreciably. Polyurethane tends to crack, craze, and split upon drying after contact with ethanol [3.7]. Threaded pipe connections are best sealed by using teflon tape, since ethanol dissolves most other pipe dopes developed for gasoline and diesel fuel.

The fuel lines onboard flexible fuel vehicles using ethanol will typically be designed to accommodate methanol fuels and should be more than adequate for ethanol. Most fuel system components designed for gasoline are likely also to be compatible with ethanol. In a test of a 1994 model fuel injected vehicle, only slight stiffening of the fuel line was observed [3.11]. No other materials compatibility problems were observed in the fuel system.

Little is known about fuel transfer hose compatibility with ethanol. Some of the hoses used for gasoline may be adequate. Experience has demonstrated that dispensing hose made for gasoline will tolerate gasoline that has 10 volume percent ethanol [3.10]. Suppliers of fuel hose should be consulted when choosing hose that will be used for transferring ethanol.

Like methanol, the lubricating oil additive package should be tailored to be compatible with ethanol to prevent leaching of the additives from the oil.

Natural Gas

Natural gas by itself is very benign and raises few materials compatibility problems. Steel is frequently used for natural gas pipelines and mains while plastic lines are typically used to bring natural gas into residences where pressures are very low. Natural gas contains two contaminants that cause most of its materials compatibility concerns: water vapor and hydrogen sulfide. Condensed water vapor will attract particulates and foreign matter and the resulting solution is often ionic and conducive to causing corrosion of most metals. Hydrogen sulfide is particularly corrosive to metals in the presence of water. If the amount of water vapor can be kept below the lowest dew point temperature that is likely to be experienced, corrosion due to hydrogen sulfide can be controlled [3.12]. If condensed water is present, carbon dioxide may also present corrosion possibilities by causing the water to become acidic. Very small amounts of methanol may be present in natural gas, added in pipelines to act as an antifreeze for condensed water. Methanol is quite corrosive to aluminum and magnesium in particular, and other metals in certain situations. When natural gas is compressed or lique-fied, the effects of contaminants on corrosion are drastically altered, as explained in the following sections.

Compressed Natural Gas (CNG)

When natural gas is compressed for use in vehicles, several materials compatibil-ity problems can arise that are not encountered in typical natural gas systems. Compressors can put significant amounts of lubricating oil into the CNG which can foul regulators and other devices where clearances are small. (The industry is developing compressors that put very little oil into the CNG, called "oil-less" compressors.) Some compressors also use methanol injection to prevent con-densed water from freezing in the system. This is more of a problem for CNG systems since, as the pressure of CNG is reduced from the maximum storage pressure (i.e., up to 4000 psi in the storage cylinders), the temperature can drop dramatically in the localized area where the pressure drop is occurring such as in a pressure regulator or in a fitting that has a high pressure drop.

CNG fuel systems can thus be subjected to compressor oil and condensed water vapor that would not occur otherwise if the gas had not been compressed. The compressor oil is probably more of an operating problem than a materials com-patibility problem, though new CNG vehicle fuel systems that use multi-point fuel injectors may encounter some problems.

A more serious threat to the materials compatibility of CNG fuel systems is con-densed water vapor. Water can cause steel and cast iron to rust and aluminum to corrode. Any corrosion of components that must withstand high pressures is a concern, since corrosion stress cracking can occur which can result in failure of the component with disastrous results. The presence of water greatly accelerates the corrosion properties of the hydrogen sulfide that might be found in the natu-ral gas. For these reasons it has been recommended that the way to control corro-sion in CNG systems is to remove sufficient water vapor to prevent it from condensing in the system under static conditions [3.13]. Natural gas dryers have been developed to help reach this goal.

85

Even if water vapor does not condense in CNG systems under static condition, some water vapor may condense in the portions of the system where pressure is reduced. The combination of water vapor and sulfur compounds has been known to cause the formation of hydrates, which are crystalline in structure (similar to snow) and which can cause operational and materials compatibility problems. Ways to prevent hydrate formation include limiting water vapor and sulfur in the natural gas, and through good system design.

Liquefied Natural Gas

Liquefied natural gas (LNG) is a very clean fuel in that the few impurities present in natural gas (water vapor, hydrogen sulfide, carbon dioxide, particulates and foreign matter, etc.) are removed almost completely when the natural gas is liquefied. Liquefaction also removes most of the hydrocarbons heavier than propane so that the resulting fuel is 95-99% methane with the remainder being primarily ethane with a smaller amount of propane. The liquefaction process is so efficient at removing contaminants that it removes the odorant placed in natural gas for transmission over pipelines (natural gas odorants are primarily mercaptans which contain sulfur and are relatively large molecules).

Being very clean and pure minimizes the materials compatibility concerns for LNG. However, LNG presents a new materials compatibility concern: operation at cryogenic temperatures.[3] For LNG fuel tanks, stainless steel is the preferred material and instances of materials compatibility problems are rare. Aluminum also has been used as a tank material without materials compatibility problems. Carbon steels are not used since their performance at low temperatures is questionable, i.e., they become susceptible to brittle fractures. While tanks are usually made from stainless steel or aluminum, LNG fittings may use some nickel alloys, brass, and copper, in addition to stainless steel and aluminum.

Few elastomers are relied on in LNG fuel systems because of the cryogenic temperatures. Welded fittings, compression fittings, and flanged joints are commonly used with little need for elastomers for sealing purposes. For the places where elastomers must be used, teflon is the preferred choice since it retains

---

[3] There is no strict definition of cryogenic, but a general consensus exists that liquids with boiling points of −100°C (−148°F) or less are cryogenic liquids.

much of its suppleness unlike typical synthetic elastomers that become brittle at low temperatures.

The above discussion about LNG materials compatibility focuses on the LNG fuel tank and portions of the LNG fuel system that are subjected to cryogenic temperatures. In LNG fuel systems there is a portion of the system downstream in which only LNG vapor at low pressures is present. This portion of the LNG fuel system is not subject to the demands placed on materials for service at cryogenic temperatures.

LP Gas

LP gas is predominantly propane which will drive its materials compatibility properties. Propane itself is fairly benign, and most propane tanks are made from commonly available and inexpensive steels or from aluminum. Propane tanks must be built to U.S. Department of Transportation (DOT) Regulations or American Society of Mechanical Engineers (ASME) *Boiler and Pressure Vessel Code* or *Code for Unfired Pressure Vessels for Petroleum Liquids and Gases* [3.14]. Given that safety relief valves for propane tanks release at pressures of 1723 kPa (250 psi) and 2151 kPa (312 psi) (for externally mounted and internally mounted tanks, respectively), these pressure vessel codes provide a margin of safety of four to five times the burst strength of the tank. Internal or external corrosion will subtract from this safety margin. The maximum vapor pressure of propane is 1482 kPa (215 psi) at 38°C (100°F), and pressures approaching 1723 kPa (250 psi) have been observed.

Propane is a good solvent for other hydrocarbons and the plasticizers[4] used in elastomers. It is important that the right fuel hoses be used for propane. Hoses made from butyl rubber are not compatible with propane and will swell and leak. Hoses made from nitrile or neoprene should be used and are compatible with LP gas. However, even in hoses made from compatible materials, residues may result from these hoses primarily from two sources: grease inside the hose from manufacturing and plasticizers present in the hose that are "leached out" by the propane upon first contact. (The extent of leach-out can be a function of the

---

[4] Plasticizers is a generic term used to describe components in elastomers that are added to make them more supple and deformable. The composition of plasticizers is not fixed and, in most cases, is proprietary for a given elastomer.

quality of the LP gas.) This conditions the hose, and subsequent plasticizer removal is small though it continues over time. The fact that plasticizers are leached out does not indicate a defective hose; it is a common occurrence and the same situation exists with dispenser hoses made for other fuels.

Impurities in propane can include water, particulates and foreign matter, and residue which includes dissolved components that are left behind when propane is evaporated. Propane residue can come directly from the refinery or gas processing plant where propane is produced, but more commonly it is picked up as propane is distributed. Sources of the residue include pipelines and tanks that contained gasoline or diesel fuel and that have not been cleaned thoroughly. Propane residue is primarily higher hydrocarbons but may include mercaptan (odorizer) and other materials soluble in hydrocarbons.

Propane is the only alternative fuel other than liquefied natural gas or liquefied hydrogen to enter the fuel tank as a liquid and enter the engine as a completely vaporized gas. This makes propane very susceptible to any contaminants that have higher boiling points, such as higher hydrocarbons. Contaminants may be introduced in the production process, during transport, and during storage. One common source of contamination is transport via multi-product pipeline when propane follows diesel fuel or other distillate fuel. The propane readily dissolves the higher hydrocarbons and carries them along, but when the propane is vaporized (predominantly in the converter in vehicle applications) the higher hydrocarbons do not vaporize and are left behind to accumulate. Only material with boiling points higher than the temperature of the converter (i.e., 82-88°C; 180-190°F) will be left behind unvaporized. An interesting characteristic of this residue is that it typically has a very strong sulfur-like odor which is believed to come from the mercaptan in the propane being absorbed into the residue as propane continually passes by it. As the mercaptans concentrate in the residue, their impact on materials compatibility increases. The propane residue may be corrosive to metal fittings and it has been observed to cause propane vapor hose to become permeable over time. The propane industry recognizes this problem and is experimenting with and developing vehicle fuel system filters and dispensing system filters to minimize the amount of residue that ends up in propane systems and the effects it can have on materials compatibility.

Hosing is typically used to transfer propane after it has been vaporized, e.g., between the vaporizer and the mixer unit of the vehicular propane fuel system.

The manufacturer of the hose must certify that it is compatible with propane as both a liquid or a vapor and must include reinforcing wire made from stainless steel [3.14].

Propane is moved (in both liquid and gaseous forms) using piping or tubing. Piping for propane can be wrought iron or steel (black or galvanized), brass, copper, or polyethylene. Tubing can be made from steel, brass, copper, or polyethylene. If the piping or tubing is subjected to the downstream of a pressure relief valve, it must be capable of withstanding the pressure rating of that relief valve (typically 1.74-2.43 MPa [250-350 psi]). In applications where higher pressures are experienced, the piping or tubing must have pressure ratings at least equal to those pressures [3.14].

The valves used in propane systems must be made from steel, ductile (nodular) iron, malleable iron, or brass. Soft parts of these valves such as gaskets, valve seat disks, packing, seals, and diaphragms must be made of materials that are certified by the manufacturer to be compatible with propane. Valve pressure ratings shall be consistent with the pressures observed in the intended application [3.14].

Flanged connections used in propane systems shall have gaskets that are compatible with propane and also have a melting point of greater than 816°C (1500°F) [3.14]. Whenever flanged joints are opened in propane systems, the gasket is to be replaced.

For other propane equipment such as pumps, pressure regulators, vaporizers, meters, strainers, compressors, etc., aluminum and zinc are acceptable materials in addition to the materials for piping, tubing, and valves.

Vegetable Oils

Physically, vegetable oils and their methyl and ethyl esters are very similar to diesel fuel. There has been no indication that any of the metals currently used in the distribution, storage, dispensing, or onboard fuel systems for diesel fuel would not be compatible with vegetable oil fuels. However, there are reports of some signs of incompatibilities with fuel transfer hoses [3.15], and nitrile and butadiene elastomers [3.16] with methyl esters. Elastomers with high fluorine content have not exhibited any problems to date. More testing is needed to fully define the elastomers that are best for use with vegetable oils.

Hydrogen

Hydrogen molecules are the lightest and smallest of all the fuel molecules. This poses unique materials compatibility concerns for both metals and nonmetals.

Hydrogen will be transmitted and stored as either a gas or cryogenic liquid. Long-term exposure of hydrogen to carbon steel can cause hydrogen "embrittlement" which makes the steel more susceptible to stress fractures. It is believed that the hydrogen molecules migrate within the steel structure and promote stress corrosion cracks to grow to the point that failure occurs. Areas most prone to hydrogen embrittlement include portions of steel parts that may have undergone strain-hardening such as bends in piping or where piping has been welded together. High-strength steel alloys are more susceptible to embrittlement than low-strength alloys. Hydrogen embrittlement is increased when hydrogen is stored under pressure, when the purity of the hydrogen is high, and when temperatures are near ambient (surprisingly, hydrogen embrittlement is not a problem with liquid hydrogen) [3.17]. At temperatures above about 220°C (430°F) hydrogen tends to permeate steel and react with the iron carbide to form methane. The steel loses strength and the methane can cause cracks to be initiated. Hydrogen embrittlement in steel that has been corroded is accelerated because hydrogen can permeate the steel more quickly [3.18]. Non-ferrous metals are not significantly impacted by hydrogen.

Metals that remain ductile at very low temperatures are preferred for use with liquid hydrogen. Examples include aluminum, copper, Monel, Inconel, titanium, austenitic stainless steels, brass, and bronze [3.19].

Hydrogen is compatible with virtually all elastomers and does not pose materials compatibility concerns. However, hoses used for hydrogen must be designed to be very non-porous to prevent permeability of hydrogen through the walls.

References

3.1. SAE Cooperative Research Program Project Group 3, "A Discussion of M85 (85% Methanol) Fuel Specifications and Their Significance," SAE Report No. CRP-2, Society of Automotive Engineers, Warrendale, Pa., September 1991.

3.2. Brinkman, N.D., *et al.*, "General Motors Specifications for Fuel Methanol and Ethanol," General Motors NAO Research and Development Center, Warren, Mich., November 1993.

3.3. Personal communication with Mr. Earl Cox, Ford Motor Company, October 1995.

3.4. Ecklund, E.E., and Bechtold, R., "Methanol Vehicles," *Encyclopedia of Energy Technology and the Environment*, John Wiley & Sons, Inc., 1995.

3.5. American Petroleum Institute, "Alcohols and Ethers—A Technical Assessment of Their Application as Fuels and Fuel Components," API Publication 4261, Second Ed., July 1988.

3.6. Cowart, J.S., *et al.*, "Powertrain Development of the 1996 Ford Flexible Fuel Taurus," SAE Paper No. 952751, Society of Automotive Engineers, Warrendale, Pa., 1996.

3.7. Mueller Associates, Inc., "Status of Alcohol Fuels Utilization Technology for Highway Transportation: A 1986 Perspective, Volume I—Spark Ignition Engines," Report No. ORNL/Sub/85-22007/4, National Technical Information Service, Springfield, Va.

3.8. Chamberlin, W.B., and Gordon, C.L., "Methanol-Capable Vehicle Development: Meeting the Challenge in the Crankcase," SAE Paper No. 902152, Society of Automotive Engineers, Warrendale, Pa., 1990.

3.9. Bechtold, R.L., Yelne, A.J., and Laughlin, M.L., "New York State Thruway Authority Alternative Fuel Vehicle Demonstration," Final Report Task 19, prepared for the New York State Energy Research and Development Authority, 1996.

3.10. Archer Daniels Midland Co., "Fuel Ethanol—Technical Bulletin," Decatur, Ill., September 1993.

3.11. Jones, B., *et al.*, "A Comparative Analysis of Ethanol Versus Gasoline as a Fuel in Production Four-Stroke Cycle Automotive Engines," SAE Paper No. 952749, Society of Automotive Engineers, Warrendale, Pa., 1995.

3.12. Lyle, F.E., "Effects of Natural Gas Contaminants on Stress Corrosion of Compressed Natural Gas Fuel Storage Cylinders," Paper #98, The NACE Annual Convention and Corrosion Show, March 1991.

3.13. "Recommended Practice for Compressed Natural Gas Vehicle Fuel," SAE J1616, Society of Automotive Engineers, Warrendale, Pa., February 1994.

3.14. National Fire Protection Association, "NFPA 58—Standard for the Storage and Handling of Liquefied Petroleum Gases," 1995 Ed., Quincy, Mass., February 7, 1995.

3.15. Spataru, A., and Romig, C., "Emissions and Performance from Blends of Soya and Canola Methyl Esters with ARB#2 Diesel in a DDC 6V92TA Engine," SAE Paper No. 952388, Society of Automotive Engineers, Warrendale, Pa., 1995.

3.16. Korbitz, W., "Status and Development of Biodiesel Production and Projects in Europe," SAE Paper No. 952768, Society of Automotive Engineers, Warrendale, Pa., 1995.

3.17. Escher, W.J.D., "Technical Overview of Hydrogen Production and Delivery Technologies," Escher:Foster Technology Associates, Inc., St. Johns, Mich., December 1980.

3.18. Treseder, R.S., "Guarding Against Hydrogen Embrittlement," *Chemical Engineering*, June 21, 1981.

3.19. Bowen, T.L., "Hazards Associated with Hydrogen Fuels," 11th IECEC.

Chapter Four

Storage and Dispensing

Storing and dispensing gasoline and diesel fuel does not present much in the way of engineering challenges since neither imposes significant constraints on materials and both are stored at atmospheric pressure. The favored method of storage for both gasoline and diesel fuel has been underground tanks made of steel or fiberglass. Fuel is pumped from the tanks using either a submerged pump at the bottom of the tank, or a pump inside the dispenser. Service stations with multiple dispensers tend to use submerged pumps for reasons of cost, though flow rate can be affected when multiple dispensers are in use at one time. Figure 4-1 illustrates a typical underground fuel storage and dispensing system for gasoline or diesel fuel.

Underground storage of gasoline and diesel fuel has several advantages: The fuel stays at a relatively constant temperature; very little above-ground space is required (only that needed for the dispenser); and refilling can be done using gravity feed rather than using a pump. However, the Achilles heel of underground tanks is leaks. Depending on the soil conditions (wet or dry, acid or base), steel tanks will rust through at varying speed with the result that fuel is lost which contaminates the local ground water supply. Such leaks are difficult to detect since they tend to be constant and represent only a small fraction of the total amount of fuel being dispensed (though over time the leaks accumulate to a large quantity). Also, since the amount of fuel in the tank is constantly changing, noticing small leaks is very difficult unless meticulous records are kept.

The U.S. Environmental Protection Agency (EPA) established regulations for underground storage tanks (USTs) containing petroleum fuels in 1988. These regulations went into effect immediately for any new USTs, and regulations

ELEMENTS OF AN UNDERGROUND PIPING SYSTEM

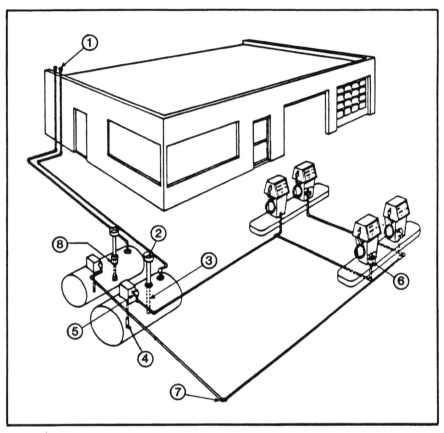

**A TYPICAL PIPING SYSTEM AT AN UNDERGROUND STORAGE
FACILITY, SHOWING THE FOLLOWING ITEMS:**

① STORAGE TANK VENT
② TANK FILL CAP AND DROP TUBE ADAPTER
③ TANK DROP TUBE
④ SUBMERGED TRANSFER PUMP
⑤ PIPELINE LEAK DETECTOR
⑥ EMERGENCY SHUTOFF VALVE
⑦ SWING JOINT AT A CHANGE OF DIRECTION
⑧ EXTRACTOR ASSEMBLY (FOOT VALVE)

*Fig. 4-1 Illustration of a Typical Underground Fuel Storage and Dispensing Facility.
(Source: Ref. [4.11])*

applying to existing USTs were phased in over a 10-year period. The proposed regulations created four minimum requirements for all new USTs [4.1]:

1. The owner or operator must certify that the UST is installed properly.
2. The UST must be protected from corrosion. A steel UST must be "cathodically" protected, as described above, and coated with a corrosion-resistant coating. Other USTs must be made totally of a noncorrodible material or of a composite of steel and noncorrodible material.
3. The UST must be equipped with devices that prevent spills and overfills. Also, correct tank filling practices must be followed.
4. The UST must have a leak detection method that provides monitoring for leaks at least every 30 days.

Several leak detection methods can be used with USTs. These methods are listed as follows and are illustrated in Figure 4-2.

- Tank tightness testing (twice yearly) and inventory control (measured daily).
- Automatic monitoring of product level and inventory control.
- Monitoring for vapors in the soil.
- Monitoring for liquids in the ground water.
- Monitoring an interception barrier.
- Interstitial monitoring within secondary containment.
- Other methods as approved by EPA.

USTs that were already in the ground when the regulations became effective must incorporate the following three improvements over a period of 10 years:

1. They must meet the same requirements for corrosion protection that apply to new USTs.
2. They must meet the new UST requirements for having a leak detection system.
3. They must be equipped with devices that prevent spills and overfills.

The UST regulations are being administered by each state. The state plans may differ in some respects to the general provisions listed in the previous paragraphs. USTs containing hazardous materials must have secondary containment. Of the alternative fuels covered in this book, methanol and ethanol must be stored in USTs having secondary containment. (Propane does not require secondary

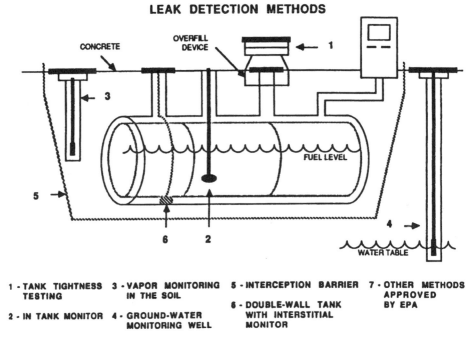

Fig. 4-2 UST Leak Detection Methods Acceptable to EPA. (Source: Ref. [4.1])

containment because it leaks as a gas that does not contaminate the soil or water supply. It is unknown whether pure biodiesel would have to be stored in USTs with secondary containment, but if mixed with diesel fuel such as B20, it would have to be.) The only USTs not covered by the regulations are farm and residential tanks with less than 1100 gallons of petroleum motor fuel used for non-commercial purposes.

Corroded tanks are not the only source of underground leaks. The EPA estimates that leaks from piping joints and fittings are more frequent than UST fuel leaks. Leaks from joints and fittings are usually due to faulty installation (not tightening joints properly or not applying sealant are two common examples) or from components that are cracked or damaged during installation.

Most states are also developing regulations for above-ground storage tanks used for petroleum fuels and hazardous materials. These regulations typically include provisions for secondary containment, overfill protection, and spill protection.

The Alcohols

The physical properties of methanol and ethanol present storage challenges very similar to gasoline and diesel fuel. Both are similar in density and have vapor pressures that are equal to gasoline or less. The primary difference is that methanol and ethanol have different effects on materials than gasoline or diesel fuel do. Much of the same storage tank and dispensing technology developed for gasoline and diesel fuel can be applied to methanol and ethanol.

Methanol

Storing and dispensing methanol presents similar engineering challenges compared to storing and dispensing gasoline. However, there are several differences that must be accounted for when designing storage and dispensing systems. Though methanol has a density similar to gasoline, it is a highly polar liquid that is completely miscible with water. This greatly limits the materials that are compatible with methanol. Stainless steel is the best material for methanol tanks, pipes, and components, but it is very expensive compared to most other materials used for fuel tanks and piping. Plain steel is used by the methanol industry and works fine as long as the methanol does not contain water or ions that will increase its corrosiveness. Aluminum cannot be used unless it is nickel-plated because methanol corrodes aluminum; the aluminum oxide components will plug filters in vehicle fuel systems and adversely affect fuel injector operation [4.2]. But even nickel plating is not foolproof since a small scratch in the plating can set up its own corrosion cell, and corrosion can proceed underneath the plating at a rapid pace. Anodized aluminum has shown better resistance than plain aluminum, but it will eventually corrode like aluminum.

Methanol is also very aggressive to most elastomers. The elastomer least affected by methanol is teflon, though elastomers with high fluorine content have been shown to be acceptable.

Figure 4-3 illustrates a schematic of a methanol service station using underground tanks. It is very similar to that required for gasoline, except that any underground piping must be double-walled. Figure 4-4 illustrates a schematic of an above-ground methanol refueling system. This particular system uses a dispenser mounted on the ground next to the tank. Dispensers that mount directly in the front or top of the tank are also available.

Tanks

Methanol tanks can be made from stainless steel, carbon steel, or fiberglass (using resins compatible with methanol). The steel tanks are welded together while the fiberglass tanks are molded around a form with the ends "glued" on. Some steel tanks are coated internally with an epoxy to prevent corrosion over time. The epoxies made for gasoline are not compatible with methanol and will come off the walls of the tank within a short time, causing major problems with

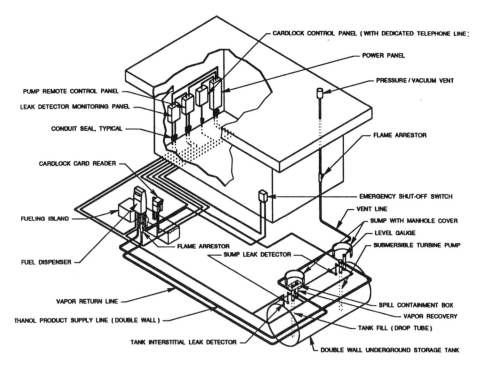

Fig. 4-3 Schematic Drawing of a Commercial Methanol Refueling System.
(Source: Ref. [4.12])

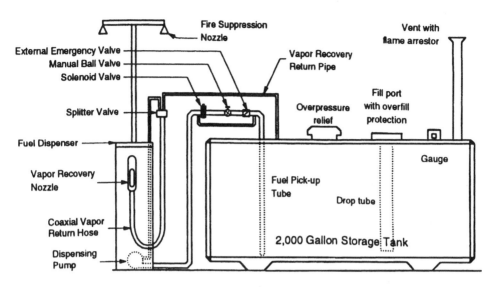

Fig. 4-4 Schematic of Above-Ground Methanol Refueling Facility. (Source: Ref. [4.13])

pumps and filters. Steel tanks using methanol do not really need tank linings since water bottoms do not occur, but methanol-compatible linings are available.

Methanol tanks that are placed underground must have secondary containment because methanol is classified as a hazardous chemical by the EPA [4.1]. Secondary containment includes:

- Double-wall tanks
- Placing the tank in a concrete vault
- Lining the excavation area surrounding the tank with natural or synthetic liners that cannot be penetrated by the chemical

Methanol is hygroscopic and will draw water vapor from the air. To minimize water absorption and vapor emissions, conservation vents are recommended for tanks holding methanol. Conservation vents are usually configured to allow venting only when the pressure in the tank exceeds 7-21 kPa (1-3 psi), and when the vacuum in the tank exceeds 5-10 cm (2-4 in.) of water. (Conservation vents are available with a wide range of pressure and vacuum settings.) By delaying venting and ingress of air, exposure of the methanol to water vapor is minimized.

Since the vapor space in methanol tanks will be flammable more frequently than for gasoline (neat methanol only—M85 should not present a difference in comparison to gasoline), flame arresters are recommended in the vent line from the tank. Most conservation vents can be ordered with a flame arrester built in, or they can be added to the vent line in addition.

Conservation vents and flame arresters should not be made from aluminum since aluminum corrodes faster when in contact with methanol vapors than with liquid methanol. Corroded conservation vents and flame arresters will contaminate the methanol and possibly not work properly.

Piping

Piping for methanol can be stainless steel, carbon steel, or fiberglass. When using carbon steel, it is important to use only black steel and not galvanized steel. Methanol will readily strip the galvanizing from the steel and the debris will collect in the dispenser and vehicle filters and cause operational problems. Pipes can be connected using threads, welding, or solvents (fiberglass).

Because methanol is classed as a hazardous chemical by the EPA, all piping placed underground must have secondary containment. Piping above-ground need not have secondary containment.

For threaded pipe connections, the preferred pipe dope is teflon paste or teflon tape. Almost all pipe dope developed for use with gasoline will be dissolved by methanol, creating leaks. Bolted connections using gaskets should use a gasket that is compatible with methanol such as teflon. In some cases, paper-type gaskets may also be acceptable if it has been proven by extensive testing that they are compatible and leak-tight.

Dispensers

Dispensers for gasoline and diesel fuel typically use steel, cast iron, aluminum, brass, bronze, and sometimes stainless steel components. Of these, only the steel, cast iron, and stainless steel components are compatible with methanol. The other parts will need to be nickel-plated or substitutes made from a compatible metal. Gasboy, Tokheim, and Wayne-Dresser have produced methanol-compatible versions of their gasoline dispensers.

Most dispensers use spin-on type filters, similar in appearance and construction to most common engine oil filters. While most use steel housings, the paper filter element and the glue used to fasten the element to the housing are typically not compatible with methanol. The most durable methanol dispenser filters to date have used nylon filter elements and methanol-compatible glue. Even with these compatible components, I have observed pitting of the steel housing of used methanol filters.

Because methanol is very aggressive to materials and the products of corrosion can cause problems in methanol vehicle fuel systems, it is recommended that the filter elements used in methanol dispensers be 3-μm mean diameter, instead of the 10-μm filters common for gasoline. Finer filter elements may become obstructed more quickly than the typical gasoline filters, reducing dispenser flow rate as they become more obstructed. How quickly filters become obstructed is mainly a function of fuel handling practices. When a refueling system is new, it is not uncommon for the dispenser filter to become obstructed after only a few hundred gallons of methanol have been pumped through it. This is because the filter very quickly catches all the debris that is typically left over from assembly (small metal chips from assembling threaded pipe; dirt; pieces of pipe dope or teflon tape; gum and residue from interior surfaces if any of the components were in gasoline service previously).

Filters with fine elements such as described for methanol are more susceptible to build-up of static electricity than typical filter elements that have larger mean diameter holes. The static electricity builds up on the filter element, which usually is not a good conductor of electricity. In severe cases, the discharge of the static electricity from the element to the housing can cause rapid erosion of the housing from the inside, eventually causing a hole to appear. Replacement of the filters at an interval that does not cause significant obstruction should minimize the chances of excessive static electric build-up.

Dispensing hoses made for gasoline absolutely <u>must not</u> be used for methanol fuel. Gasoline dispensing hose will rapidly deteriorate and put material into the vehicle that will in turn quickly obstruct the onboard fuel filter. Goodyear has developed methanol-compatible dispensing hose which is easily recognized since it is green or blue in color. It is coaxial, making it compatible with vapor recovery systems. Though this hose is compatible, Ford Motor Co. has found it to release plasticizers when methanol is first put through it, which can obstruct vehicle fuel filters (because they are 1 μm) [4.3]. Ford's recommendation is to soak the hose in methanol for 24 hours before putting it into use (the methanol used to soak the hose will need to be disposed of properly).

Like dispensing hoses, nozzles must be made specifically for methanol fuel. Gasoline nozzles usually contain aluminum parts and elastomers that are not compatible with methanol. Most gasoline nozzles will function for a time, but even when they are working, they are putting corrosion products directly into the vehicle. Emco-Wheaton makes a dispensing nozzle compatible with methanol and configured for vapor recovery systems.

Miscellaneous Components

Breakaway fittings for dispensing hoses are frequently required to guard against the person who drives his vehicle away from the dispenser before removing the nozzle from the filler tube on the vehicle. Breakaway fittings are available in stainless steel which is compatible with methanol.

For suction pump systems where the potential for syphoning exists should the fuel feed line be broken when the dispenser is in operation, some jurisdictions require a valve that allows fuel flow from the tank only when the dispenser is in operation. These valves are typically made from brass and have an elastomeric diaphragm used to open the valve. The diaphragm within the valve must be methanol-compatible or it will rupture, causing the pump to draw air and lose prime. The valves should be nickel-plated and the diaphragm replaced with one made from teflon [4.4].

Leak Detection Systems

Passive types of leak detection such as observation wells and collection sumps where product is collected and analyzed directly should work effectively with methanol. Active leak detection systems that rely on thermal conductivity and electrical resistivity sensors will not work with methanol because its properties are so different from gasoline. Another type of active leak detection system that will work with methanol or any other type of fuel relies on changes in impedance in a sensor wire as it becomes wetted with the fuel [4.5]. These leak detection systems also have the advantage that they can pinpoint the location of the leak along the length of the sensor wire.

Fire Suppression Systems

Many states now require fire suppression systems for public service stations and for some private fleet refueling facilities. Most of these systems are dry chemical using infrared detectors and work as effectively on methanol vehicle fires as gasoline vehicle fires. No special requirements are necessary for these systems when used at a methanol refueling facility. All fire suppression systems should be checked after installation for proper coverage using a fraction of the dry chemical that would normally be used (this is called a "puff test"). The puff test not only shows the coverage that will be obtained, but tests the components of the fire protection system.

Lightning Protection

All methanol refueling systems located outdoors should take lightning protection into account. Lightning strikes can damage the refueling system and potentially cause fires. Generic building codes are available for lightning protection (NFPA 180) which should be followed and supplemented by local building code requirements, if any.

Ethanol

Compared to storing and dispensing methanol, little research has been conducted to determine the optimum materials compatibility of ethanol with refueling system components. As a consequence, most installations of ethanol storage and dispensing systems have relied on the research conducted for methanol. As the demand for ethanol storage and dispensing systems grows, components optimized for ethanol will no doubt begin to appear.

Ethanol presents the same engineering challenges as methanol compared to storing and dispensing gasoline. Like methanol, ethanol is completely miscible with water, and the presence of water (especially if it is highly ionic from road salt or other contaminants) can make ethanol much more corrosive than it would be in its pure state. It is also possible that ethanol will contain some acids from production. Carbon steel tanks will work fine with ethanol fuel as long as it does not contain ionic impurities that increase its corrosiveness. Stainless steel is the best material for ethanol tanks, pipes, and components, but it is very expensive compared to most other materials used for fuel tanks and piping. Unlike methanol, aluminum can be used with ethanol which should make adaptation of several components much easier than for methanol. However, like carbon steel, ethanol with a high water content and ionic impurities may have sufficient corrosivity to initiate corrosion of aluminum. Under these circumstances, the zinc coating of galvanized steel may also be affected. Bronze appears to be unaffected by ethanol fuel.

Ethanol is more aggressive to most elastomers than gasoline. The elastomer least affected by ethanol is teflon, though elastomers with high fluorine content have been demonstrated to be acceptable. Other elastomers can be used with ethanol if they have been proven to be compatible.

Tanks

Ethanol tanks can be made from stainless steel, carbon steel, or fiberglass (using resins compatible with ethanol). The steel tanks are welded together while the fiberglass tanks are molded around a form with the ends "glued" on. Some steel tanks are coated internally with an epoxy to prevent corrosion over time. The epoxies made for gasoline are not compatible with ethanol and will come off the walls of the tank within a short time, causing major problems with pumps and filters. Steel tanks using ethanol do not really need tank linings since water bottoms do not occur, but ethanol-compatible linings are available.

Ethanol is hygroscopic and will draw water vapor from the air. To minimize water absorption and vapor emissions, conservation vents are recommended for tanks holding ethanol. Conservation vents are usually configured to allow venting only when the pressure in the tank exceeds 7-21 kPa (1-3 psi), and when the vacuum in the tank exceeds 5-10 cm (2-4 in.) of water. (Conservation vents are available with a wide range of pressure and vacuum settings.) By delaying venting and ingress of air, exposure of the ethanol to water vapor is minimized.

Since the vapor space in ethanol tanks will be flammable more frequently than for gasoline (neat ethanol only—E85 should not present a difference in comparison to gasoline), flame arresters are recommended in the vent line from the tank. Most conservation vents can be ordered with flame arresters built in, or they can be added to the vent line in addition. Unlike methanol, conservation vents made from aluminum should not experience corrosion from ethanol vapors and should be acceptable.

Piping

Piping for ethanol can be stainless steel, carbon steel, or fiberglass. Ethanol should not attack the zinc in galvanized steel if the fuel has low water content and is not contaminated by ions. Ethanol not meeting these requirements can strip the galvanizing from the steel, and the debris will collect in the dispenser and vehicle filters and cause operational problems. Pipes can be connected using threads, welding, or solvents (fiberglass).

For threaded pipe connections, the preferred pipe dope is teflon paste or teflon tape, but some of the pipe dopes developed for use with gasoline may also be compatible with ethanol. (Contact the manufacturer to verify that compatibility is proven, not just suggested.) Bolted connections using gaskets should use a gasket that is compatible with ethanol. Many of the commonly used elastomers are compatible with ethanol, but if there is any doubt, teflon is sure to work. Cork gaskets are always to be avoided when using ethanol fuel because ethanol deteriorates cork rapidly. Paper-type gaskets may also be acceptable if it has been proven by extensive testing that they are compatible and leak-tight.

Dispensers

Dispensers for gasoline and diesel fuel typically use steel, cast iron, aluminum, brass, bronze, and sometimes stainless steel components. Ethanol should not pose a problem with any of these metals if the ethanol has low water content and few ionic impurities. To be safe, dispensers configured for methanol should prove to be completely adequate for ethanol.

Most dispensers use spin-on type filters, similar in appearance and construction to most common engine oil filters. Most use steel housings and paper filter element that are compatible with ethanol, but the glue used to fasten the element to the housing may not be compatible with ethanol. The filter manufacturer should be contacted to see whether they have tested their filters for compatibility with ethanol (a recommendation without testing verification should be looked upon skeptically). If there is doubt, dispenser filters having nylon filter elements and ethanol-compatible glue should prove compatible.

The vehicles designed to use ethanol as a fuel tend to have very fine filters to protect their fuel injectors and other components from corrosion products.

Because of this, it is recommended that the filter elements used in ethanol dispensers be at least 3-μm mean diameter, instead of the 10-μm filters common for gasoline. Finer filter elements may become obstructed more quickly than the typical gasoline filters, reducing dispenser flow rate as they become more obstructed. How quickly filters become obstructed is mainly a function of fuel handling practices. When a refueling system is new, it is not uncommon for the dispenser filter to become obstructed after only a few hundred gallons of ethanol have been pumped through it. This is because the filter very quickly catches all the debris created that is typical from assembly (small metal chips from assembling threaded pipe; dirt; pieces of pipe dope or teflon tape; gum and residue from interior surfaces if any of the components were in gasoline service previously).

Filters with fine elements such as described for ethanol are more susceptible to build-up of static electricity than typical filter elements that have larger mean diameter holes. The static electricity builds up on the filter element, which usually is not a good conductor of electricity. In severe cases, the discharge of the static electricity from the element to the housing can cause rapid erosion of the housing from the inside, eventually causing a hole to appear. Replacement of the filters at an interval that does not cause significant obstruction should minimize the chances of excessive static electric build-up.

Dispensing hoses made for gasoline absolutely <u>must not</u> be used for ethanol fuel. Gasoline dispensing hose will rapidly deteriorate and put material into the vehicle that will in turn quickly obstruct the onboard fuel filter. Goodyear has developed dispensing hose that is compatible with methanol which is easily recognized since it is green or blue in color, and should be compatible with ethanol. It is coaxial, making it compatible with vapor recovery systems. As when using methanol, it is recommended that this hose be soaked in ethanol for 24 hours before putting it into use to remove any plasticizers that may be dissolved by ethanol (the ethanol used to soak the hose will need to be disposed of properly).

Unlike dispensing hoses, nozzles made for gasoline may be compatible with ethanol fuel. The metals contained in dispensing nozzles should all be compatible with ethanol, but it is not guaranteed that all the elastomers will be. Consult the nozzle manufacturer to obtain their recommendation.

Leak Detection Systems

Passive types of leak detection such as observation wells and collection sumps where product is collected and analyzed directly should work effectively with ethanol. Active leak detection systems that rely on thermal conductivity and electrical resistivity sensors will not work with ethanol because its properties are so different from gasoline. Another type of active leak detection system that will work with ethanol or any other type of fuel relies on changes in impedance in a sensor wire as it becomes wetted with the fuel [4.5]. These leak detection systems also have the advantage that they can pinpoint the location of the leak along the length of the sensor wire.

Fire Suppression Systems

Many states now require fire suppression systems for public service stations and for some private fleet refueling facilities. Most of these systems are dry chemical using infrared detectors and work as effectively on ethanol vehicle fires as gasoline vehicle fires. No special requirements are necessary for these systems when used at an ethanol refueling facility. All fire suppression systems should be checked after installation for proper coverage using a fraction of the dry chemical that would normally be used (this is called a "puff test"). The puff test not only shows the coverage that will be obtained, but tests the components of the fire protection system.

Lightning Protection

All ethanol refueling systems located outdoors should take lightning protection into account. Lightning strikes can damage the refueling system and potentially cause fires. Generic building codes are available for lightning protection (NFPA 180) which should be followed and supplemented by local building code requirements, if any.

Natural Gas

To use natural gas as a vehicle fuel it must either be compressed or liquefied so that sufficient amounts of energy can be stored onboard to give the vehicle a practical operating range.[1] Natural gas refueling facilities must be located away from buildings, public streets and sidewalks, railroad tracks, other refueling dispensers, property lines, and public places such as parks or parking lots. The following sections describe the refueling system components for both compressed natural gas (CNG) and liquefied natural gas (LNG).

Compressed Natural Gas

A natural gas compression system will always include the basic elements of a compressor, some sort of compressed gas storage, and a means of dispensing the CNG into vehicles.[2] There are two types of CNG refueling systems: slow-fill and fast-fill. In slow-fill systems, several vehicles are connected to the output of the compressor at one time. These vehicles are then refilled over several hours of compressor operation. In fast-fill systems, enough CNG is stored so that several vehicles can be refueled one after the other, just like refueling from a single gasoline dispenser. Figure 4-5 illustrates the basic elements of a fast-fill CNG refueling facility.

The CNG storage is divided up into several tanks called a cascade. The pressure of the CNG in the cascade is higher than the maximum storage pressure of the cylinders on the vehicle. The cascade attempts to deliver as much of its CNG to vehicles as possible before the pressure difference decreases to where the flow rate slows dramatically. This is accomplished by the refueling system control system, which can be as simple as all-mechanical control or as complex as complete computer control. No matter what the control system, most fast-fill systems will be able to refuel only a certain number of vehicles before the cascade is depleted, after which refueling will occur at the rate of the compressor output. In very large fleets where refueling occurs for several hours straight, the compressors must be sized to match the refueling demand, and a buffer storage system is included to keep the compressors operating continuously between vehicle refuelings.

---

[1] A few early natural gas vehicles have used low pressure storage of natural gas in large bags stored on the roof (in the case of buses) or carried behind the vehicle on a trailer. This approach is not practical for most modern vehicles due to height and volume constraints, and since trailers would interfere severely with operations.

[2] CNG can also be made directly from LNG, obviating the need for a compressor.

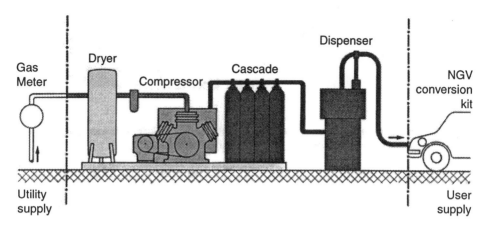

*Fig. 4-5 Schematic of the Basic Elements of a Fast-Fill CNG Refueling System.
(Source: Sulzer Technology Corporation)*

Most CNG refueling systems also should include a dryer that removes water vapor, foreign matter, and hydrogen sulfide from the natural gas before it is compressed. The water vapor can condense in the vehicle fuel system, causing corrosion, especially if hydrogen sulfide is present. In some parts of the country where the natural gas is known to be consistently very clean, a dryer may not be necessary, but in most cases it is prudent to include one.

Fast-fill CNG refueling systems need a dispenser to direct the CNG into the vehicles. Dispensers include a meter to measure the amount of the CNG that went into the vehicle and, for public refueling stations, the cost of that gas. In slow-fill systems, a hose and a connector is provided for each vehicle, and metering to determine the amount of CNG going into the vehicle is usually cost prohibitive. For this reason, slow-fill is usually considered practical only for vehicles of a single fleet.

Compressors

Compressors for natural gas refueling systems are typically reciprocating compressors powered by electric motors or by natural gas engines. Depending on the refueling system, the output of the compressor can be as high as 31 MPa (4500 psi). Natural gas is typically delivered via pipeline at a pressure below 35 kPa (5 psi). To achieve the high pressures desired requires that the gas be compressed through multiple sequential stages, usually four. (The gas is cooled between each stage to reduce the work of compression and ensure that it does not become dangerously warm.) The valves and rings of compressors need periodic maintenance to ensure that they work properly and so that the efficiency and capacity of the compressor is not compromised.

The materials in compressors are primarily cast iron, steel, and aluminum. The quality of the natural gas in the distribution system is such that materials compatibility problems are few. If a dryer is included in the system, materials compatibility concerns should be essentially non-existent.

Storage

Since weight is not a concern for CNG storage cylinders used in CNG refueling systems, these cylinders are made almost exclusively from carbon steel. The regulations to which these cylinders are made include the U.S. DOT and the ASME pressure vessel code. The cylinders made to DOT regulations are extruded from sections of forged billets while the ASME cylinders are typically welded together. ASME cylinders are preferred for refueling system storage because they are larger in volume and minimize the amount of high-pressure plumbing needed to connect them with the compressor and dispensing system (see Figure 4-6).

CNG storage cylinders are typically placed on a concrete slab without enclosure, though a canopy over cascades of DOT cylinders is useful in protecting the cylinder valves from the elements. ASME cylinders are placed horizontally in groups of three arranged vertically, and typically do not have shut-off valves that need protection from canopies or enclosures. CNG storage cylinders can be placed in underground vaults, though this is uncommon and presents some additional safety concerns such as an asphyxiation hazard to personnel in the vault in the presence of a large leak. Vaulted CNG storage facilities would most likely require explosion-proof wiring and lighting throughout, adding to its cost. Protection from ignition sources above the vault vent(s) is a concern and would have to be addressed.

Fig. 4-6 A CNG Cascade Using Three ASME Cylinders.

Piping

Piping for CNG refueling facilities must be compatible with natural gas and must be capable of four times the rated service pressure without failure. Stainless steel seamless tubing is most commonly used for CNG piping, and piping made from plastic, galvanized steel, aluminum, or copper alloys (where the copper content exceeds 70%) is not allowed. Threaded and compression-type fittings that do not use gaskets or sealants are allowed. Threaded pipe must use a compatible pipe dope; threaded pipe may not be used underground [4.6].

Dispensers

Dispensers for CNG provide a convenient means of directing CNG from the storage system into the vehicle. They will typically incorporate some sort of on-off switch activated by removal of the refueling nozzle and a meter to measure the amount of CNG dispensed. A typical CNG dispenser as used at commercial CNG refueling facilities is shown in Figure 4-7.

CNG dispensers typically incorporate stainless steel for the flowpath of the natural gas, and no materials compatibility problems are likely. CNG dispensers do not typically incorporate filters—if filtering is deemed necessary, it is usually done at the outlet of the compressor before the CNG enters the storage system.

Fig. 4-7 A Commercial CNG Dispenser.

CNG dispensing nozzles are made from aluminum and stainless steel and no materials compatibility problems have been noted. The mechanical action of fastening the CNG nozzle appears to be the limiting factor in CNG nozzle life, not deterioration due to corrosion from the natural gas. CNG dispensing nozzles have a finite lifetime typically characterized by the total number of refuelings completed. Nozzles should be replaced at the end of their useful life to prevent inadvertent failure.

CNG dispensing hose uses an impervious inner tube reinforced by a braided over-wrap of stainless steel wire. The inner tube can be teflon or a similar material that is not porous and will not become overly stiff in cold weather. An unusual problem that can occur with high-pressure hose is that if static electricity builds up inside the hose, it may discharge through the hose to the wire reinforcement. When this occurs, a small hole is formed in the inner tube that quickly leads to a leak through the wire reinforcement. Such incidents are rare, but they do occur, caused by the non-conducting properties of the inner tube. There have been no reported incidents of such failures in CNG dispensing hoses.

Like CNG dispensing nozzles, CNG dispensing hoses have finite lifetimes—often measured in the number of refuelings completed. When the lifetime of the hose has been exceeded, it should be replaced to prevent inadvertent failure during use. Hose manufacturers will have guidelines for determining when their hose should be replaced.

Control Systems

The control systems for CNG refueling systems have evolved from being only mechanically controlled to some that are completely computer controlled today. Computer control offers flexibility not possible with mechanical systems and the ability to change the control strategy of the refueling system. Computer-controlled systems can also provide functions such as accounting of the amount of fuel dispensed into vehicles and billing functions. It is also possible to incorporate safety shut-down of the system if certain conditions are present.

Leak Detection Systems

All CNG refueling systems should have some type of leak detection system using methane detectors. Methane detectors should be placed strategically about the refueling area at points near probable release sites, and they will need to be placed inside compressor enclosures (if they are used). Since methane rises when released, methane detectors should be placed above ground level at points where released gas is likely to pass. The methane detectors provide outputs to activate alarms when certain levels of methane are detected. Most detectors are set up to alarm when 20% of the lower flammability limit (LFL) of methane in air is detected (this is about 1% methane in air). Some refueling facilities have used a two-stage approach to methane detection and alarm. When 20% LFL is detected, a first level alarm is sounded. If the concentration increases, say to 40% LFL, additional measures such as shut-down of the refueling system is initiated. All refueling systems should have manual shut-down switches located both right in the vicinity of the refueling system and at a location remote from the refueling system so that, should there be an existing emergency, shut-down can be accomplished from a safe distance. The manual shut-down switches should be wired so that they can be reset only if the methane detectors are not in alarm, i.e., the concentration they are sensing is less than 20% LFL.

Fire Suppression Systems

Many states now require fire suppression systems for public service stations and for some private fleet refueling facilities. Most of these systems are dry chemical using infrared detectors and work as effectively on natural gas fires as gasoline fires. No special requirements are necessary for these systems when used at a CNG refueling facility. All fire suppression systems should be checked after installation for proper coverage using a fraction of the dry chemical that would normally be used (this is called a "puff test"). The puff test not only shows the coverage that will be obtained, but tests the components of the fire protection system.

Lightning Protection

All CNG refueling systems located outdoors should take lightning protection into account. Lightning strikes can damage the refueling system and cause fuel release and/or fires. Generic building codes are available for lightning protection (NFPA 180) which should be followed and supplemented by local building code requirements, if any.

Liquefied Natural Gas

There are several transportation applications where natural gas storage as CNG provides inadequate range because of limited space and allowable weight for a CNG fuel system. The railroad industry and heavy-duty over-the-road trucks are the two current primary examples of transportation vehicles that as a practical matter must use LNG instead of CNG in order to use natural gas as a fuel.

LNG will either be made on-site at a refueling facility by liquefying pipeline natural gas or delivered on-site already liquefied. In both cases, LNG will need to be stored in bulk at the refueling facility and equipment will be necessary to dispense the LNG into vehicles. Moving the LNG from the bulk storage tank to the dispenser can be accomplished using a pump or, if the pressure in the bulk tank is high enough, by "pressure transfer." Using a pump to transfer the LNG has the advantage of higher output pressures and the ability to keep the pressure in the bulk tank low. The disadvantage of pump transfer is that pumps fail and they represent a source of heat gain to the LNG causing vapor generation. Pressure transfer is simpler but problems may arise should the pressure in the vehicle tank be higher than the bulk tank pressure, which is very possible since vehicle LNG tanks are often made for higher operating pressures than bulk LNG storage tanks[3]. An additional consideration in making the decision to use pump or pressure transfer is that should a pump fail, a pressure transfer of LNG can always be made.

Figure 4-8 shows an LNG refueling facility designed for refueling LNG transit buses. This facility includes a dispenser, kiosk, canopy, methane detection system, fire suppression system, and trailer-mounted LNG storage tank. In this system, LNG is pumped from the storage tank to the dispenser and then on to the bus. A trailer-mounted tank was chosen for this application since a permanent tank was not desired.

---

[3] Bulk LNG storage tanks typically have maximum operating pressures of less than 0.86 MPa (125 psi) to keep costs low (higher operating pressures require thicker storage tank walls), while LNG storage tanks on vehicles may have operating pressures as high as 1.72 MPa (250 psi). The higher operating pressure of vehicle LNG tanks is justified because heat gains tend to be larger per unit volume, the additional weight due to thicker tank walls is not as critical, and any vapor vented is lost completely at a high cost to the vehicle operator.

Fig. 4-8 The Maryland MTA LNG Refueling Facility.

Storage

LNG tanks are all double-walled because an evacuated space is required to achieve sufficient insulation to keep the LNG from vaporizing too quickly. A "hold time" (time from when the tank is filled to when pressure build-up requires vapor to be vented) of at least five days is required for practical LNG tanks. In addition to the evacuated interstitial space, several different types of insulation may be used. The goal is to make the conductive, convective, and radiative heat transfer into the LNG as low as possible.

LNG tanks are made from stainless steel or aluminum according to ASME pressure vessel codes. Since LNG is a very clean fuel (no water, very little foreign matter, and no sulfur-based odorant), there are no problems with materials compatibility. The primary concern for LNG tanks is the thermal cycling they must endure from ambient temperature to −162°C (−260°F).

LNG tanks are most often cylindrical but may also be spherical. The center section of cylindrical tanks is made from rolled sheet welded together. The ends are usually formed under hydraulic pressure and then welded to the center section. Stand-offs are welded between the inner and outer tanks to connect them and provide stability in operation.

LNG tanks are designed for fairly low pressures, typically less than 1.72 MPa (250 psi), otherwise their advantage over storing natural gas as CNG would be lost. As LNG warms, vapor is produced that builds up pressure in the tank. To protect the tank from rupture due to over-pressure, vent valves are included that will release some vapor when the maximum set pressure is reached. Vent valves come in two types: those that reseat when the pressure in the tank falls below the maximum, and those that don't. The second type is used only to protect the tank from over-pressure rupture, while the first type is used to control the pressure in the tank. Often they are used together in the same tank. Both types of vent valves may be made from stainless steel, brass, and aluminum; materials compatibility problems are rare because of the cleanliness of LNG.

To date, LNG tanks have all been designed and installed as above-ground tanks. With the desire to use LNG as a vehicle fuel comes the desire to be able to install LNG tanks underground. The reasons to install LNG tanks underground include: reduced space required at refueling facilities; no need for spill containment dikes; ability to place LNG refueling facilities in densely populated urban areas; no need for secondary containment systems since LNG is non-toxic and lighter than air when vaporized; reduced potential for vandalism; greater protection of the tank from fires; and greater public acceptance [4.7]. Underground LNG tanks must be protected from corrosion, and actions must be taken to prevent the surrounding ground from freezing which could put external stress on the tank. One of the technical impediments still facing underground LNG tanks is design of a suitable in-tank pump to lift the LNG from the tank to the dispenser. Work is ongoing to design and demonstrate underground storage of LNG.

Piping

Piping for LNG comes in two types: single-wall and vacuum-jacketed. All the single-wall piping is made from stainless steel for durability purposes at cryogenic temperatures. The inner section of vacuum-jacketed piping is also stainless steel for the same purpose. Single-wall piping may also be insulated through the addition of external polyurethane or multi-layered insulation.

Piping for LNG service is connected via threaded joints, flanged connections, swaged fittings, or welded connections. Welding provides the most secure connection and should be used whenever possible. Flanged connections are suitable and should be used where welded connections cannot. Threaded connections should be used only sparingly, unless welded after installation. Pipe threads cut into stainless steel tubing are hobbed, i.e., the tops of the threads are trimmed off. Without this precaution, stainless steel pipe will continue to cut its own threads when tightened. Hobbed threads present a challenge to making a threaded connection completely leak-tight. Pipe dope that is specific to cryogenic service should always be used with threaded stainless steel connections. (Chemically, LNG presents no materials compatibility concerns for pipe dope.) If threaded connections are used in several consecutive 90-degree turns, obtaining a leak-tight system becomes very difficult. For these reasons, threaded connections should be used only when other connections cannot. Swaged fittings work well but are limited to fairly small-diameter piping. Flanged fittings tend to loosen if they are subjected to repeated temperature cycling.

Long runs of LNG piping will expand and contract significant amounts when LNG is present. Bellows-type expansion joints are recommended to prevent stresses from building up in piping systems due to temperature cycling. These expansion joints can be welded into the piping or attached using flanged connections.

Pressure relief valves play an important role in LNG piping systems. Any section of LNG piping that can trap LNG must have a pressure relief valve to prevent failure of that section of pipe should all the LNG vaporize. These pressure relief valves are similar to those used on LNG storage tanks. Spring-loaded valves that reseat following a reduction in pressure are preferred to those that do not and vent all the LNG in that line. Pressure relief valves should be protected from water accumulation to prevent freezing in the open position since the LNG vapor that is released will likely be very cold.

Gaskets for flanged connections in cryogenic service are readily available and compatibility with LNG is not a problem.

Dispensers

LNG dispensers are very similar to CNG dispensers in function: their primary purpose is to meter and control the flow of LNG to the vehicle. Like CNG dispensers, LNG dispensers do not include a pump. The meters inside LNG dispensers are typically of the Coriolis acceleration type that have low flow restriction and use stainless steel internal flow surfaces. Piping inside the dispenser is typically vacuum-jacketed stainless steel.

LNG filters in LNG dispensers are not common, though most will have at minimum stainless steel wire mesh strainers that will keep metal shavings and large foreign matter from proceeding into the vehicle fuel tank.

LNG dispenser hoses should have breakaway fittings to prevent a vehicle from pulling away from the dispenser with them attached. Since the LNG hoses are securely fastened to the vehicle when refueling, pulling away without disconnecting the hoses will cause the hoses to be severed or the dispenser to be damaged.

The standard practice to date for personnel who refuel LNG vehicles is to require them to wear a full face shield, and cryogenic gloves and apron. As LNG refueling systems become more automated, the need for such safety equipment may be eliminated. CH•IV Cryogenics has designed an LNG refueling system for which it is claimed the refuelers need not wear any safety equipment since the refueling hoses do not have LNG in them when they are connected or disconnected [4.8]. A high level of automation of the refueling system is required to achieve this feat.

LNG refueling hose connectors are made by Parker-Hannifin and Moog, Inc. Both are "dry-break" connectors which allow LNG to flow only after engagement of the connectors.

Control Systems

LNG dispensing control systems are unique in that they must incorporate a "cool-down" cycle. The purpose of the cool-down cycle is to cool the piping down to the same temperature as the LNG. To achieve this, LNG is circulated through the dispenser and dispensing piping (the dispensing piping must be connected to the dispenser to allow LNG to flow through it) which creates a significant amount of LNG vapor that must be directed back to the storage tank. The control system must be able to open and close valves, turn on pumps, and read thermocouples throughout the process.

The cool-down cycle can be automated so that an attendant need only initiate the process and wait for the dispenser to indicate the cool-down is completed.

Control is provided either by programmable logic controllers or by using a PC-based control system.

Leak Detection Systems

All LNG refueling systems must have some type of leak detection system using methane detectors, especially since LNG has no odor to warn of leaks. Methane detectors should be placed strategically above the refueling area, and placement inside the dispenser is another possible location. The methane detectors provide outputs to activate alarms when certain levels of methane are detected. Most detectors are set up to alarm when 20% of the lower flammability limit (LFL) of methane in air is detected. (This is about 1% methane in air.) Some refueling facilities have used a two-stage approach to methane detection and alarm. When 20% LFL is detected, a first level alarm is sounded. If the concentration increases, say to 40% LFL, additional measures such as shut-down of the refueling system is initiated. All refueling systems should have manual shut-down switches located both right in the vicinity of the refueling system and at a location remote from the refueling system so that, should there be an existing emergency, shut-down can be accomplished from a safe distance. The manual shut-down switches should be wired so that they can be reset only if the methane detectors are not in alarm, i.e., the concentration they are sensing is less than 20% LFL.

Fire Suppression Systems

Many states now require fire suppression systems for public service stations and for some private fleet refueling facilities. Most of these systems are dry chemical using infrared detectors and work as effectively on natural gas fires as gasoline fires. No special requirements are necessary for these systems when used at an LNG refueling facility. All fire suppression systems should be checked after installation for proper coverage using a fraction of the dry chemical that would normally be used (this is called a "puff test"). The puff test not only shows the coverage that will be obtained, but tests the components of the fire protection system.

Small leaks of LNG will vaporize very quickly and the vapor will rise just as released CNG might. However, if the leak of LNG is large enough to cause a pool of LNG to develop, the vapors that form may be cold enough to stay near the surface in a cloud that will move sideways with any prevailing breeze. These clouds will be recognizable because the vapors will be cold enough to condense water vapor in the air, giving the cloud the appearance of fog. Any ignition source in the path of the cloud could ignite it, creating a fire.

Control of pools of spilled LNG can help to mitigate the hazards of LNG clouds. The first line of defense would be to use concrete with high insulation value in the area of the refueling dispenser. This limits the heat transfer rate to the pool and slows the rate of vapor formation. Once a pool has formed, vaporization can be further limited and controlled by covering it with fire-fighting foam. This insulates the pool from above and further limits the formation of LNG vapor. Water should never be sprayed onto LNG pools since it will cause vigorous boiling of the LNG. Containment and control of LNG pools is the preferred method for controlling LNG spills.

Lightning Protection

All LNG refueling systems located outdoors should take lightning protection into account. Lightning strikes can damage the refueling system and cause fuel release and/or fires. Generic building codes are available for lightning protection (NFPA 180) which should be followed and supplemented by local building code requirements, if any.

LNG-to-CNG Systems

LNG-to-CNG systems are of interest to those who want to use CNG as a vehicle fuel and have access to inexpensive LNG. Advantages of these systems include lower capital and operating costs, and smaller footprint compared to the compressors needed to produce the same quantity of CNG from pipeline natural gas.

These systems convert LNG to CNG by exploiting the inherent thermodynamic properties of a phase change from liquid to gas of natural gas. By relying on heat transfer from the ambient air to produce CNG, operating costs to produce CNG are reduced significantly from the costs of operating a compressor.[4] The quality

---

[4] Some LNG-to-CNG systems use natural gas heaters to provide sufficient heat to quickly vaporize the LNG.

of CNG produced from LNG should be higher than CNG from typical pipeline natural gas since the liquefaction process removes all water vapor, sulfur compounds, and a portion of the higher hydrocarbons in some cases. CNG from LNG does not have odorant in it and the system should provide a means for odorizing the CNG before it is put into vehicles.

LNG-to-CNG systems employ a combination of LNG and CNG technology plus heat exchangers not used in either type of stand-alone system. No unique materials compatibility requirements are created by these systems—the LNG portion of the system has the same materials compatibility concerns as LNG-only systems and the CNG portion follows CNG materials compatibility concerns.

LP Gas

LP gas, or propane, is the alternative fuel used in highest volume in the U.S. at present. Propane is unique among alternative fuels in that it is a pressurized liquid, i.e., a modest pressure (under 43.5 kPa [300 psi]) will maintain it in the liquid state. Propane fuel tanks must thus be built to pressure vessel codes like CNG and LNG tanks, but propane is transferred using pumps because pressure differentials are low and the pressure cannot be manipulated as it can for LNG. Propane is sometimes stored in refrigerated containers which lowers the storage pressure significantly, though such containers are rarely if ever used to store propane intended for use in vehicles.

Propane refueling facilities consist of a bulk storage tank, a transfer pump, and a dispenser for refilling vehicles (see Figure 4-9). Propane refueling facilities must be located away from buildings, public streets and sidewalks, railroad tracks, other refueling dispensers, property lines, and public places such as parks or parking lots. They must be located away from any pits, and drains or blow-off valves must not be directed to within 4.6 m (15 ft) of sewer system openings. Diking of the bulk storage tank is not required except for refrigerated propane tanks. The National Fire Protection Agency Standard 58 (NFPA 58) is the standard for propane refueling facilities most widely consulted by building code officials across the U.S.

Fig. 4-9 A Propane Refueling System with Vertical Storage Tank.
(Source: Tom Gorman Company)

Storage

Propane storage tanks must be built to either U.S. Department of Transportation (DOT) regulations for cylinders or American Society of Mechanical Engineering (ASME) pressure vessel codes. Steel is the most common material for propane tanks, though aluminum is also allowed and is popular for portable propane tanks.

The types of construction include welding, brazing, or die forming from ingots (similar to how steel CNG cylinders are formed). Propane tanks for refueling vehicles can be located above or below ground level, but there are restrictions as to how close tanks can be placed to buildings and property lines. Tanks mounted above ground should be placed on a concrete pad, or mounted within a steel frame that keeps the tank above the ground surface. Above-ground tanks should be painted a light reflecting color—white is typically used. Tanks placed underground should be protected from corrosion through the use of coatings and cathodic protection.

Bulk storage propane tanks have spring-loaded pressure relief valves to prevent over-pressure in service. The valves must be made from materials having melting points of 816°C (1500°F) or higher. Typical materials used to construct pressure relief valves include steel, ductile iron, malleable iron, and brass. Gaskets for valves must have a similar temperature rating and be made from metal. Aluminum O-rings and spiral-wound metal gaskets are also acceptable.

Piping

Both piping and tubing can be used for propane. Propane piping can be made from steel (black or galvanized), wrought iron, brass, copper, or polyethylene. Propane tubing can be made from steel, brass, copper, or polyethylene. Connections for piping and tubing can include threaded, welded, or brazed. Polyethylene shall be joined by heat fusion. Any trapped sections of pipe must include a pressure relief valve to prevent over-pressure. All piping shall be supported to limit vibration and prevent stresses from being imposed that might cause failure.

Dispensers

Dispensers for propane can be configured like dispensers for gasoline with readouts for gallons dispensed and accumulated total purchase price. Dispensers should be mounted on a concrete pad for stability and protection from corrosion, unless the dispenser is part of a complete storage and dispensing unit where the dispenser is mounted securely within the system. Dispensers cannot be mounted inside buildings, but can be mounted under canopies and can be protected from the elements as long as the area is ventilated and the perimeter is not more than 50% enclosed. The "heart" of a propane dispenser is the meter that measures the amount of propane dispensed. These meters are of the constant volume displacement type and are typically made from steel and aluminum. Dispensers also must incorporate an excess flow valve or emergency shut-off valve at the point where the dispensing hose connects to the propane piping inside the dispenser. Dispensing hoses are limited to 5.5 m (18 ft) in length and must incorporate an emergency breakaway device. Propane nozzles are similar to gasoline nozzles in appearance, but they must be securely fastened to the refueling coupling before transfer of the propane under pressure can occur.

The materials used in propane dispensers include steel, wrought iron, brass, and aluminum. Dispensing nozzles are made from aluminum, brass, and steel. Few elastomer gaskets are used in propane systems—most are aluminum or steel. Propane dispensing systems should incorporate filters to prevent debris and heavy oils from being pumped into the vehicle fuel tank.

Control Systems

Propane refueling systems do not need elaborate control systems. The dispenser controls operation of the transfer pump, and emergency shut-down switches should be mounted near the dispenser and at some location between 6.1 m (20 ft) and 30.5 m (100 ft) from the dispenser.

Leak Detection Systems

Leak detection systems are not required for propane refueling systems but could be incorporated in the vicinity of the dispenser, near ground level since propane vapors are heavier than air. Like pipeline natural gas, propane is odorized so that leaks should be noticed by refueling personnel.

Fire Suppression Systems

Propane refueling facility fire suppression systems can be dry chemical or water-based; the decision about which to use depends on the siting of the facility and on local fire protection codes and regulations. Emergency plans to deal with inadvertent releases of propane or propane fires should be worked out with the local fire, police, and emergency response agencies. Propane fires are not recommended to be extinguished until the source of the burning propane is shut off.

Lightning Protection

All propane refueling systems located outdoors should take lightning protection into account. Lightning strikes can damage the refueling system and potentially cause fires. Generic building codes are available for lightning protection (NFPA 180) which should be followed and supplemented by local building code requirements, if any.

Vegetable Oils

Vegetable oil based fuels will more than likely be esters of vegetable oils (biodiesel) that have very similar physical characteristics to typical No. 2 diesel fuel. Because of this, biodiesel can be stored and dispensed using the same equipment that is currently used for diesel fuel. The only difference of note is that biodiesel is more aggressive to the elastomers that may be used in pumps and meters. High fluorine-content elastomers have been demonstrated to work with biodiesel, though direct replacement of existing elastomers with high-fluorine-content elastomers may not always be possible. Teflon is also an acceptable replacement material for elastomers affected by biodiesel.

Storage

Biodiesel can be stored in the steel or fiberglass tanks currently used for diesel fuel. All the regulations for above-ground and underground diesel fuel tanks apply to tanks holding biodiesel. Conservation vents are not required since the vapor pressure of biodiesel is very low, as it is for diesel fuel.

Piping

Piping for biodiesel can be steel (black or galvanized), fiberglass, or plastic that is suitably rated for fuel use. Methods of connection can include threads, flanges, welding, and brazing for metal piping, and threads, heat fusion, or solvent fusion for fiberglass or plastic piping. In all cases, the joints shall be tested for leaks, and piping shall be properly supported to prevent stresses from being imposed on piping systems from installation. Little or no experience exists for the compatibility of pipe dopes with biodiesel. Experience with pipe dope used with alcohol fuels would suggest that some (albeit very slow) deterioration might be expected since biodiesel may contain very small quantities of methanol or ethanol depending on whether the biodiesel is a methyl or ethyl ester. Teflon tape can be used as a thread sealant and its compatibility with biodiesel is assured.

Dispensers

The same dispensers used for diesel fuel should work fine for biodiesel. However, dispensers do have a few elastomers and they may be affected if they do not have a high fluorine content. Close observation of joints that use elastomers should be done to determine compatibility.

At present, there is no indication that the filters used in diesel fuel dispensers now are not compatible with biodiesel.

Like elastomers, there is very little experience with the long-term compatibility of dispensing hoses and nozzles with biodiesel. (The only concern with the nozzles should be the elastomeric components used inside.) Should dispensing hoses prove not to be durable over time, those dispensing hoses developed for methanol should certainly be compatible.

Control Systems

No control system beyond a start-stop switch on the dispenser for the transfer pump is needed. An emergency shut-down switch should be installed at a location remote from the dispenser that will shut down all the electricity to the dispensing station.

Leak Detection Systems

Underground storage tanks and piping should incorporate a leak detection system. While biodiesel is biodegradable, fuel additives or diesel fuel may be added to biodiesel which will make spills a health concern if they find their way into the water system. There is no need for leak detection around the dispenser.

Fire Suppression Systems

Many states now require fire suppression systems for public service stations and for some private fleet refueling facilities. Most of these systems are dry chemical using infrared detectors and work as effectively on biodiesel fires as on gasoline or diesel fuel fires. No special requirements are necessary for these systems when used at a biodiesel refueling facility. All fire suppression systems should be checked after installation for proper coverage using a fraction of the dry chemical that would normally be used (a "puff test"). The puff test not only shows the coverage that will be obtained, but tests the components of the fire protection system.

Lightning Protection

All biodiesel refueling systems located outdoors should take lightning protection into account. Lightning strikes can damage the refueling system and potentially cause fires. Generic building codes are available for lightning protection (NFPA 180) which should be followed and supplemented by local building code requirements, if any.

Hydrogen

While hydrogen is acknowledged to be an excellent alternative fuel due to its low emissions and capability for production from renewable sources, very few vehicles have ever been built that were powered by hydrogen. The U.S. Department of Energy and BMW have experimented with passenger cars using liquid hydrogen [4.9, 4.10]. Even when liquefied, hydrogen has only 20% of the energy contained in the same volume of gasoline. Others have experimented using compressed hydrogen onboard as the fuel source, even though it has only 5% the energy of gasoline per unit volume. Vehicles powered by fuel cells using hydrogen might be able to use compressed hydrogen because the high efficiency of fuel cells will reduce the amount of fuel required to achieve a practical operating range.

Storage

Liquid hydrogen must be stored in highly insulated containers similar to those used for liquefied natural gas. These containers are made from stainless steel and may be spherical or cylindrical in shape. The containers are double-walled to allow use of vacuum insulation in addition to insulation to prevent heat transfer from conduction, convection, and radiative sources. Pressure relief devices are incorporated to release hydrogen should the internal pressure of the tank increase to the maximum safe operating limit.

The containers used to store hydrogen as a compressed gas are similar in construction to containers for compressed natural gas. These containers can be made from steel, aluminum, or composite materials. They must incorporate pressure relief devices so that the hydrogen is vented in the event that the tank is exposed to fire.

Piping

Stainless steel piping is preferred for transferring hydrogen. Liquid hydrogen requires use of vacuum-jacketed piping to minimize heat gain. Compression fittings and flanged joints are preferred to threaded connections since the threads on stainless steel pipe must be hobbed, i.e., the tops of the threads trimmed off. Without this precaution, stainless steel pipe will continue to cut its own threads when tightened. Hobbed threads present a challenge to making a threaded connection completely leak-tight. (Chemically, hydrogen presents no materials compatibility concerns for pipe dope. Pipe dope that is specific to cryogenic service should always be used with threaded stainless steel connections where liquid hydrogen may be present.) If threaded connections are used in several consecutive 90-degree turns, obtaining a leak-tight system becomes very difficult. For these reasons, threaded connections should be used only when other connections cannot be used. Flanged joints do not need to use gaskets made of special materials.

Dispensers

Only a few experimental dispensers have been built for liquid hydrogen [4.10]. Like dispensers for liquefied natural gas, these dispensers will need to be able to control cool-down of the dispenser and fuel transfer lines, and direct vaporized hydrogen back to the storage tanks or to some other means of recovery. Extensive use of stainless steel is probable in the fuel meter, and measurement of delivered liquid hydrogen and returned gaseous hydrogen is desired. The fuel transfer lines within the dispenser are likely to be vacuum-jacketed to minimize heat gain to the liquid hydrogen.

Dispensers for compressed hydrogen are likely to be very similar in design and function to dispensers for compressed natural gas.

Control Systems

Hydrogen dispensers will likely make use of programmable logic controllers or PC-based control systems. Both liquid hydrogen and compressed hydrogen dispensers will need to be able to monitor pressures and temperatures, open and close valves, turn on pumps, and interface with fire and emergency shut-down systems.

Leak Detection Systems

All hydrogen refueling systems should have leak detection since hydrogen is odorless and colorless. Hydrogen detectors should be placed strategically above the refueling area and in other appropriate locations. The hydrogen detectors should alarm when the concentration of hydrogen reaches about 1% volume in air, which is 20-25% of the lower flammability limit of hydrogen. If the hydrogen concentration continues to rise, the detectors should have the capability to shut down the refueling system until the hydrogen is removed.

Fire Suppression Systems

All refueling systems of significant size should consider the incorporation of a fire suppression system. The current systems using dry chemical should be effective on hydrogen fires, though fire suppression professionals should be consulted when choosing which type of system to use. Special sensors may be required since the infrared detectors used for petroleum fuels may not sense hydrogen fires as efficiently.

Sources of Additional Information

For more information about methanol refueling facility equipment and suppliers, contact:

- The American Methanol Institute
  800 Connecticut Avenue, N.W., Suite 620
  Washington, D.C. 20006
  Phone: 202-467-5050

- Methanol Fuel Specifications and Lists of Methanol-Compatible Refueling
  Facility Equipment
  American Automobile Manufacturers Association
  7430 Second Avenue, Suite 300
  Detroit, Michigan 48202
  Phone: 313-872-4311
  Fax: 313-872-5400

- *Methanol Fueling System Installation and Maintenance Manual*, March 1996
  The California Energy Commission
  Transportation Technology & Fuels
  1516 Ninth Street, MS-41
  Sacramento, California 95814
  Phone: 916-654-4634
  Internet Address: http://www.energy.ca.gov

For more information about ethanol refueling facility equipment and suppliers, contact:

- National Ethanol Vehicle Coalition
  1648 Highway 179
  Jefferson City, Missouri 65109
  Phone: 314-635-8445

- Ethanol Fuel Specifications and Lists of Ethanol-Compatible Refueling
  Facility Equipment
  American Automobile Manufacturers Association
  7430 Second Avenue, Suite 300
  Detroit, Michigan 48202
  Phone: 313-872-4311
  Fax: 313-872-5400

- *Guidebook for Handling, Storing, & Dispensing Fuel Ethanol*
  U.S. Department of Energy
  Alternative Fuels Data Center
  P.O. Box 12316
  Arlington, Virginia 22209
  Phone: 800-423-1DOE
  Internet Address: http://www.afdc.doe.gov

For more information about CNG and LNG refueling facility equipment and suppliers, contact:

- Natural Gas Vehicle Coalition
  1515 Wilson Boulevard, Suite 1030
  Arlington, Virginia 22209
  Phone: 703-527-3022
  Fax: 703-527-3025

- American Gas Association
  1515 Wilson Boulevard, Suite 1030
  Arlington, Virginia 22209
  Phone: 703-841-8000

- Gas Research Institute
  8600 West Bryn Mawr Avenue
  Chicago, Illinois 60631-3562
  Phone: 312-399-8100
  Fax: 312-399-8170
  Internet Address: http://www.gri.org

- Zeus Development Corporation (specializing in LNG)
  10333 Richmond Avenue, Suite 400
  Houston, Texas 77042
  Phone: 713-952-9500
  Fax: 713-782-9594

For more information about propane refueling facility equipment and suppliers, contact:

- National Propane Gas Association
  1600 Eisenhower Lane
  Lisle, Illinois 60532
  Phone: 708-515-0600
  Fax: 708-515-8774

- Propane Vehicle Council
  1101 17th Street, N.W., Suite 1004
  Washington, D.C. 20036
  Phone: 202-530-0479
  Fax: 202-466-7205

For more information about vegetable oil refueling facility equipment and suppliers, contact:

- American Biofuels Association
  1925 North Lynn Street, Suite 1050
  Arlington, Virginia 22209
  Phone: 703-522-3392

- Biofuels America
  RD 1, Box 19
  Westerlo, New York 12193-9801
  Phone: 518-797-3377

- National BioDiesel Board
  2405 Grand, Suite 700
  Kansas City, Missouri 64108
  Phone: 816-474-9407

- National SoyDiesel Development Board
  P.O. Box 194898
  Jefferson City, Missouri 65110-4898
  Phone: 800-841-5849

References

4.1. *Proposed Regulations for Underground Storage Tanks: What's in The Pipeline*, Publication Number 26A, U.S. Environmental Protection Agency, Office of Underground Storage Tanks, Washington, D.C., April 1987.

4.2. Brinkman, N.D., Halsall, R., Jorgensen, S.W., and Kirwan, J.E., "The Development of Improved Fuel Specifications for Methanol (M85) and Ethanol (E_d85)," SAE Paper 940764, Society of Automotive Engineers, Warrendale, Pa., 1994.

4.3. Personal communication with Mr. Earl Cox, Ford Motor Company, October 1995.

4.4. Bechtold, R.L., Yelne, A.J., and Laughlin, M.L., "New York State Thruway Authority Alternative Fuel Vehicle Demonstration," Final Report Task 19 - 1996, Prepared for the New York State Energy Research and Development Authority.

4.5. PermAmert Environmental Specialty Products, Inc., 7720 North Lehigh Avenue, Niles, Illinois 60714-3491.

4.6. National Fire Protection Association, *NFPA 52 - Standard for Compressed Natural Gas (CNG) Vehicular Fuel Systems, 1995 Edition*, Quincy, Mass.

4.7. Midgett, Dan E. II, "Best Available Practices for LNG Fueling of Fleet Vehicles," Gas Research Institute Topical Report No. GRI-96/0180, February 1996.

4.8. Press release, CH•IV Cryogenics, 599 Canal Street, Lawrence, Mass., August 30, 1996.

4.9. Stewart, W.F., "Operating Experience With a Liquid-Hydrogen Fueled Buick and Refueling System," *International Journal of Hydrogen Energy*, Vol. 9, No. 6, pp. 525-538, 1984.

4.10. Birch, S., "Advanced Research at BMW," *Automotive Engineering*, Vol. 102, No. 10, October 1994, Society of Automotive Engineers, Warrendale, Pa.

4.11. "Recommended Practices for Underground Storage of Petroleum," New York State Department of Environmental Conservation, Albany, N.Y., 1985.

4.12. *Methanol Fueling System Installation and Maintenance Manual*, California Energy Commission, March 1996.

4.13. Yelne, A., Bechtold, R.L., Moog, C.G., and Laughlin, M.D., "NYSERDA AFV-FDP M85 Flexible Fuel Vehicle Fleet Operating Experience," SAE Paper No. 962068, Society of Automotive Engineers, Warrendale, Pa., 1996.

Chapter Five

Refueling Facility Installation and Garage Facility Modifications

Building Codes

It is important to know and realize that building codes are recommendations about how facilities should be constructed (e.g., refueling and garage facilities). They attempt to cover as many circumstances as possible to be comprehensive. But building codes become law only when they are adopted by the local authority having jurisdiction, which is usually the local fire marshall.

Four nationally recognized building code organizations in the U.S. cover refueling and garage facilities: the National Fire Protection Association (NFPA); the International Fire Code Institute (IFCI); the Building Officials and Code Administration (BOCA); and Southern Building Code Congress International (SBCCI). The most pertinent NFPA codes include NFPA 30—Flammable and Combustible Liquids Code, and NFPA 30A—Automotive and Marine Service Station Code. The IFCI publishes the Uniform Fire Code which is predominant in the western part of the country. Chapter 23 of the National Fire Prevention Code published by BOCA applies to storage, handling, and processing of all hazardous liquids. The SBCCI has adopted NFPA 30A virtually verbatim for fuel dispensing at non-retail facilities [5.1].

Each of these building code organizations uses different means to develop their codes and keep them current. Each one of them publishes updates on a continual basis and major revisions of their entire code periodically (every three years for the NFPA and SBCCI). The local jurisdiction having authority to enforce building codes may or may not adopt the revisions that are published between major revisions of the entire codes. It is important to find out who the local jurisdiction having authority is before beginning design of any facility, what building codes they have adopted, and the specific version of the building codes they have adopted.

Other than for propane and CNG refueling systems, it is unlikely that local jurisdictions will have building codes specific to alternative fuels. In the absence of specific codes, local officials often fall back on the codes developed for gasoline and diesel fuel as a starting point. The objective of this chapter is to present information that could be useful in educating local officials in the building code and safety issues involving alternative fuels and to provide what information is available for each. By working with local building and fire officials <u>before</u> beginning design and construction of alternative fuel refueling facilities, it is likely that some reasonable compromise can be worked out. Failure to consult them may result in having to make expensive and possibly inappropriate modifications later.

All vehicle fuels raise safety concerns. These are mainly associated with the physiological properties of the fuel, e.g., toxicity, and with its flammable properties. These risks are primarily controlled by preventing the release of fuel from vehicles, storage tanks, and during dispensing into vehicles. While fuel releases due to leaks or component failure are not predictable, their occurrence should be anticipated and hazardous situations avoided through good design, personnel training, and commitment to safe operating practices. During the vehicle repair process there are times when fuel releases will occur under controlled conditions. The hazards presented by these fuel releases can be controlled by:

- Limiting workers' exposure to released fuels (if the fuels present a physiological hazard).

- Keeping released fuel vapor concentrations outside of their limits of flammability.

- Isolating ignition sources from locations where ignitable fuel mixtures may exist.

Methods used to accomplish these goals are in a large part dependent on the physical properties of the fuel in use.

In addition to safe alternative fuel storage, dispensing, and building designs, a comprehensive personnel safety training program is essential to prevent hazardous situations from occurring during vehicle refueling, maintenance, and storage. Personnel should be informed about the properties of the alternative fuel, and safe practices for operating, refueling, and maintaining alternative fuel vehicles. A comprehensive training program combined with sound design of facilities should result in alternative fuel vehicle use having a similar or better safety record than conventional fuel vehicles.

The following sections identify several building codes that are pertinent to the placement of alternative fuel dispensing systems, fuel storage to support these systems, and codes for buildings that store and maintain alternative fuel vehicles. The discussions presented are not intended to be comprehensive—they are provided to alert you to the important codes and regulations that should be considered. Also, the discussion of the codes in the following is not a substitute for acquiring and reading the codes themselves. Building codes contain much detailed information that cannot be completely explained here. Many facilities also have unique situations which may not be covered explicitly in the codes. For these reasons, it is imperative that those wishing to install alternative fuel refueling systems, and store and maintain alternative fuel vehicles indoors, do a comprehensive analysis of their situation, and research the pertinent codes in detail before applying for building permits and initiating construction drawings.

The Alcohols

Specific codes or regulations for vehicular storage and dispensing systems for alcohol fuels do not exist. Alcohols are flammable liquids which are covered in NFPA 30, *Flammable and Combustible Liquids Code* [5.2]. Gasoline is included under NFPA 30 as a Class IA flammable liquid, while diesel fuel is included as a combustible liquid. When alcohols are used as fuel for vehicles, they almost always have some gasoline or high vapor pressure hydrocarbons in them which gives them safety characteristics similar to gasoline, and makes them Class IA flammable liquids. When methanol or ethanol are used as fuel without gasoline addition (such as in alcohol compression-ignition engines), they would fall under the Class IB flammable liquid definition. Regardless of whether significant amounts of gasoline are added to the alcohol, it is likely that local code officials would apply NFPA 30A, *Automobile and Marine Service Station Code* [5.3] to storage and dispensing systems for alcohol fuels, just as they would for gasoline or diesel fuel.

The following presents information from NFPA 30A that is pertinent to the basics of placing alcohol fuel storage and dispensing systems. What follows is not a comprehensive summary of NFPA 30A requirements—just guidelines as to the major requirements. States or local jurisdictions may have requirements that are more stringent than those of NFPA.

Location of Storage Tanks

Where storage tanks can be placed depends on: whether they are above ground or underground; the nearest building, property line, or public way; the type of organization that uses the tank and how it is used; and for above-ground tanks, whether they are fire resistant or not.

Above-Ground Tanks

According to NFPA 30A, above-ground tanks are limited to 45,600 L (12,000 gal) capacity each, and multiple tanks manifolded together cannot exceed 152,000 L (40,000 gal). Above-ground tanks must be placed at least 15 m (50 ft) from the nearest important building or public way, and at least 30 m (100 ft) from any property line. These distances can be reduced by half if the above-ground tank is fire resistant. Where the tanks are used to refuel vehicles associated with non-public operations (commercial, industrial, government, or manufacturing), no minimum distances are required if the tanks are fire resistant or placed in vaults.

Tanks to store fuel above ground must be built as above-ground tanks; tanks built as underground tanks cannot be used as above-ground tanks.

Underground Tanks

NFPA 30 provides guidance for placement of underground tanks. According to NFPA 30, underground tanks containing alcohol fuels must be placed at least 0.3 m (1 ft) from the wall of any basement or pit. They also must be at least 0.9 m (3 ft) from any property line. States have very specific regulations in addition to these that include corrosion protection, leak detection, and anchoring of underground storage tanks. (See Chapter 4.) EPA underground storage tank regulations specify that underground storage tanks and their lines must have secondary containment.

Location and Installation of Fuel Dispensers

NFPA 30A does not specify limits on the location of fuel dispensers except that they must be at least 6 m (20 ft) from any fixed source of ignition. Dispensers should be solidly mounted on a refueling island or some similar equivalent means. Protection should be provided in the form of bollards or guardrail to prevent errant vehicles from colliding with the dispenser. For above-ground tanks, dispensers can be mounted directly on the tanks.

Dispensing hoses are limited to 5.5 m (18 ft) and must have a breakaway device in the event a vehicle drives off without removing the refueling nozzle.

Vehicle Storage and Maintenance Facilities

Facilities that store and maintain vehicles using alcohol fuels should be built to the same codes as those for vehicles using gasoline or diesel fuel. Alcohol vapors from released fuel are heavier than air similar to gasoline vapor, and represent similar fire safety hazards. In addition, most alcohol fuels have significant amounts of gasoline in them, and the gasoline vapor released from spills of these fuels will cause similar fire safety hazards as gasoline alone. The ventilation rates recommended for conventional fuels should be sufficient for alcohol fuels.

Fire Protection

Fire protection may or may not be required according to local codes; facilities that serve the public are increasingly being required to provide some type of fire suppression. Regardless of whether the fuel storage and dispensing system is required to have a fire suppression system, its addition should be considered in light of personnel safety and potential for property loss.

Dry chemical fire suppression systems are the most appropriate for use at vehicle fuel storage and dispensing systems. The discharge nozzles should be placed both overhead and at curb level so that the entire area prescribed by the radius of the dispensing hose will be covered. Testing of the nozzle placement should be done by using a "puff test" in which the dry chemical is discharged on purpose to verify coverage. (The puff test need not discharge the entire amount of dry chemical, only enough so that the area of coverage can be determined with certainty.) The puff test is triggered by creating a small controlled fire of the fuel used which also tests the fire detection sensors, usually ultraviolet and infrared.

Natural Gas

There are many building codes and regulations for using natural gas as fuel for home heating and for industrial uses, but few codes and regulations exist for using natural gas as a vehicle fuel. What codes and regulations do exist focus on vehicle fuel systems, CNG refueling facilities, and LNG storage facilities. Work is ongoing to define codes, regulations, and standard operating practices for storage of CNG and storage and maintenance of natural gas vehicles indoors.

Compressed Natural Gas

NFPA 52 is the primary source of codes and regulations for installation of CNG fuel systems on vehicles and the design, construction, and operation of CNG storage and dispensing systems [5.4]. While NFPA 52 provides guidance for design and installation of CNG refueling facilities, it does not cover codes for buildings where CNG vehicles are maintained or stored.

Location of Storage Tanks

Storage of CNG containers to service vehicle refueling facilities is for practical purposes limited to outdoors because the limit for storage indoors is 283 m^3 (10,000 scf) in each building or room. Outdoor storage of CNG containers shall be above ground on stable, noncombustible foundations or in vaults that are ventilated and have drainage. CNG storage containers mounted horizontally shall be supported by not more than two points, and where flooding is possible they shall be secured to prevent floating. The CNG storage containers shall be painted to prevent corrosion, unless they are made from composites where the container manufacturer shall be contacted to determine whether painting is appropriate. CNG storage containers shall be mounted at least 3 m (10 ft) from adjoining property lines, buildings, nearest public street or sidewalk, or any source of ignition. They shall also not be located within 15 m (50 ft) of the nearest rail of any railroad main track. CNG containers shall not be located below electrical power lines or in a location that might be hit by falling electrical power lines. If containers of flammable or combustible liquids are stored nearby, the CNG containers shall be kept at least 6 m (20 ft) away, and any readily ignitable material shall not be permitted within 3 m (10 ft). Clearance of 1 m (3 ft) shall be allowed from the valves and fittings of multiple groups of storage containers to provide unimpeded access.

Location and Installation of Fuel Dispensers

CNG fuel dispensers come under the same siting requirements as CNG storage containers with the additional requirement that the location where the dispensing hose connects to the vehicle shall be 3 m (10 ft) or more from adjoining property lines, buildings, the nearest public street or sidewalk, or any source of ignition. The fill location point shall also be at least 1 m (3 ft) from any CNG storage container.

In practice, CNG dispensers are typically mounted on refueling islands, similar to those used for gasoline or diesel fuel dispensers. Bollards are recommended to prevent vehicles from inadvertently running into the CNG dispenser. If protection from the weather using a canopy is desired, the canopy shall be made from noncombustible materials and should be designed so that it will not capture natural gas released from the dispenser. Lighting provided under the canopy should be explosion-proof. Pressure relief valve lines from the dispenser shall be directed upward and terminated above the roof of the canopy. All electrical equipment for the dispenser shall be installed in accordance with NFPA 70 [5.5]. All locations within 1.5 m (5 ft) of the dispenser or any relief valve in any direction shall be classified as Class I, Division 1 (explosion-proof). The area 4.6 m (15 ft) in the direction of a pressure relief valve and 15 degrees from the centerline of discharge shall also be classified as explosion-proof for the purposes of electrical wiring.

Vehicle Storage and Maintenance Facilities

Building codes specifically addressing storage and maintenance of CNG vehicles indoors do not exist, but several engineering companies and organizations have been studying this problem and some general consensus has emerged about the hazards presented [5.6]. The primary hazard is accumulation of natural gas in a building above its lower flammability level. The simplified solution is to provide forced or natural ventilation so that this does not occur. The difficult part is to provide safety systems that will be able to handle the wide range of natural gas release rates that can occur from CNG vehicles.

Maintaining a safe environment in a vehicle repair facility is helped by:

- <u>Taking proper precautions</u> when working on CNG vehicles in a facility.

- Using facility designs that <u>minimize locations where rising natural gas can be trapped</u>.

- <u>Eliminating sources of ignition</u> from areas where natural gas may be present or travel through if released from vehicles.

- <u>Controlling ventilation rates</u> to limit volumes of combustible mixtures present in a facility after a fuel release.

Worker training and implementation of safe work procedures are one method of decreasing the hazard presented by bringing CNG vehicles into a facility. Training should emphasize methods of:

- Minimizing the possibility of a fuel release.

- Limiting the volume of fuel released should a leak occur.

- Limiting worker-related sources of ignition in locations where released fuel may be present.

- Proper response to hazardous conditions.

Even with good vehicle design and proper worker training, some failures which cause fuel to be released from vehicles will occur. Safety can be maintained by preventing ignition or accumulation of the released fuel. The ignition of the releasing or released fuel presents several hazardous scenarios.

Ignition of the leaking fuel near the source of the release will produce a flame similar to a burner or torch flame. If ignition occurs soon after the leak has begun, the initial volume of flammable mixture will be small; consequently, the danger presented will also be small. Once ignited, the main safety concern is the overheating of fuel system components or ignition of other nearby combustible materials causing the fire to spread. Ignited fuel releases should be extinguished by interrupting the flow of fuel to the fire, e.g., closing the master shut-off valve or the cylinder valves.

Far more hazardous are incidents involving ignition of accumulated quantities of flammable fuel/air mixtures. There are several ways large quantities of flammable fuel/air mixtures can be generated before ignition. Fuel leaks which release fuel at a high rate, e.g., actuation of a pressure relief device (PRD) or rupture of a main fuel line (if the cylinder valves are open), will generate a large cloud of flammable mixture (up to the equivalent of several gallons of fully vaporized gasoline) above the site of the fuel release in a short period of time. If unconfined and no sources of ignition are present, the cloud will dissipate harmlessly. If a vehicle is inside a building when a rapid fuel release occurs, the released fuel would have to be vented out of the building (by either forced or natural ventilation removing air from above the vehicle) while at the same time being kept isolated from ignition sources. Without proper ventilation, a flammable mixture will rise and form a layer under the ceiling inside the building. Ignition of such a mixture would rapidly release large amounts of energy and likely cause extensive damage from over-pressure and fire. Although leaks with high rates of fuel release should be rare, facility designs should incorporate ventilation and other features to ensure safety during these types of releases.

Another scenario where large quantities of flammable mixtures may be generated involves slow fuel leaks without ignition in locations where adequate ventilation is not provided. Slow leaks can occur at any of the fittings in the fuel system (from vibration, accident damage, or improper installation or repair), from defective regulators or faulty valves. Low rate fuel releases may be difficult to locate, depending on the rate of fuel release. Their existence will be readily known from the telltale odor, but maintenance personnel may not (if the leak is slow enough) be immediately aware of the location of the fuel leak. A small fuel leak may, over a period of time, release a large volume of natural gas. The main dangers presented by these types of leaks are associated with gas accumulation. If there is inadequate ventilation, flammable mixtures of gas can become trapped in parts of the vehicle or in the building. If the fuel continues to accumulate for a long period of time, it is possible that a large quantity of ignitable mixture may be present.

Repair areas where CNG vehicle conversions and repairs are to be performed should have provisions for either natural or forced ventilation. Current codes describe ventilation requirements only for facilities servicing conventional liquid fuel vehicles; however, these codes can be used as a basis to determine the needs of facilities servicing CNG vehicles.

Both the NFPA and the National Electrical Code (NEC) provide ventilation and electrical requirements for garages where gasoline and diesel vehicles are repaired. These standards are designed to prevent the accumulation of vapors inside the garage and limit sources of ignition in locations in which released fuel vapors are likely to be found. These NFPA and NEC requirements are illustrated in Figure 5-1 [5.7].

NFPA 88B—Standard for Repair Garages requires that areas below grade used for repair vehicles have a forced ventilation system capable of continuously removing at least 0.75 cubic feet of air per minute for each square foot (cfm/sq.ft.) of floor space [5.8]. This ventilation requirement helps prevent accumulation of heavier-than-air fuel vapors which could accumulate in below-grade areas.

NFPA 88B removes potential sources of ignition from areas where fuels may accumulate by requiring suspended unit heaters to be located at least 8 ft above the floor [5.9]. In addition, other heaters with glowing elements must be located at least 18 in. above the floor, and the garage must have continuous forced ventilation at the rate of 0.75 cfm/sq.ft. [5.10].

All electrical equipment installed less than 18 in. above the ground must comply with NEC's Class I Division 2 wiring requirements to limit the probability of ignition occurring in hazardous environments [5.11]. All electrical equipment

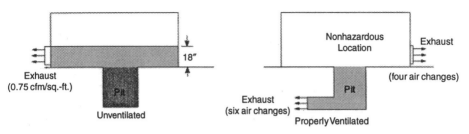

NEC Electrical Classifications

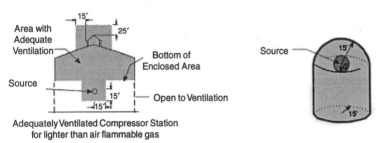

NFPA 497A Requirements   **NFPA 58 Standard for the Storage and Handling of LPG**

▓ Areas with Class 1 Divison 1 wiring requirements   ▓ Areas with Class 1 Division 2 wiring requirements

Fig. 5-1 Current NEC and NFPA Mechanical and Electrical Requirements for Both Heavier and Lighter than Air Flammable Vapors. (Source: Ref. [5.7])

installed in unventilated pits must comply with NEC's Class I Division 1 (explosion-proof) wiring requirements.

NFPA 497A—Recommended Practice for Classification of Class I Hazardous (Classified) Locations for Electrical Installations in Chemical Process Areas contains recommendations for wiring requirements in an enclosed area where lighter-than-air flammable gases may be released [5.12]. While these recommendations are for chemical processing areas, the releases are similar in the hazard they present to what would be found for natural gas released in repair garages. These recommendations are also illustrated in Figure 5-1.

NFPA 58—Standards for the Storage and Handling of Liquefied Petroleum Gases contains recommendations for areas where liquefied petroleum gases (LPGs) are handled [5.13]. Included are recommendations for wiring requirements, illustrated in Figure 5-1. While LPGs are heavier than air, the release hazards presented at ground level would be analogous to those found near the ceiling during a CNG fuel release.

Several studies have been done on the risks presented by CNG vehicles inside buildings. These studies are useful in helping to determine safety changes in the absence of clear code requirements. A recent study modeling probable fuel releases from CNG transit buses has shown that flammable mixtures *may* extend more than 18 in. down from the ceiling but that horizontal displacement of released fuel as it rises rarely exceeds 3-4 ft from the release site [5.14]. These are results of a single study, modeling various-sized fuel releases from transit buses, stored in a modern (conventional fuel designed) transit bus garage. Caution should be used before extrapolating the results of this study to other situations. For example, a PRD releases fuel quickly and with great velocity. When released under a vehicle, this high velocity fuel release may be deflected by the ground or vehicle components and travel greater distances horizontally than if it were released above a vehicle.

Fuel can accumulate to depths (distances directly below the ceiling) greater than 18 in. during periods of large-scale fuel releases. The rate at which fuel is released due to a major component failure, e.g., release by a PRD, can be quite large. One study has calculated an initial fuel release rate of 29.5 m³/min (932 scfm) when a typical vehicle PRD has activated [5.15]. A garage ventilation system removing 0.25 m³/min/m² (0.75 cfm/sq.ft.) from above a vehicle could still allow fuel to accumulate if the overall ventilation rate was insufficient to handle a 29.5 m³/min (932 scfm) release. Increasing the background ventilation rate, to account for the maximum rate of fuel release may cause problems if no fuel were being released, e.g., annoying drafts in the repair bay, need for additional heating, etc. For this reason it is recommended that gas detectors be installed at strategic locations. These gas detectors would be used to increase the ventilation rate—increasing the exhaust fan speed or activating additional exhaust fans—when released natural gas is detected. In addition to increasing the ventilation rate, the gas detectors should be wired into an alarm system to warn personnel of elevated methane concentrations above the vehicles.

Based solely on the national code requirements, a generalized work area for CNG vehicles can be developed, as shown in Figure 5-2.

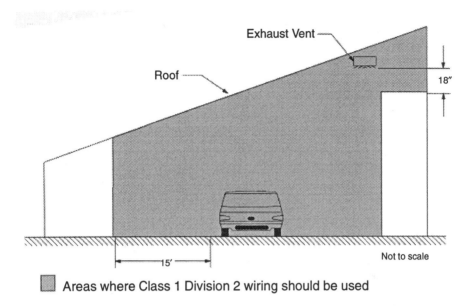

Areas where Class 1 Division 2 wiring should be used

Fig. 5-2 General Electrical Requirements for CNG Repair Areas Based on Modification of National Codes. (Source: Ref. [5.7])

The shaded areas in Figure 5-2 are based on the properties of natural gas, a review of the current garage-related national codes, and other studies of gas releases in repair facilities. The codes where these recommendations appear include:

• NFPA 497A [5.12]—All electrical systems within 15 ft of the vehicle (when in the repair bay) should be rated for use in Class 1 Division 2 locations.

• NFPA 88B [5.13]—The garage ventilation system should remove at least 0.75 cfm/sq. ft. of floor space. <u>The pick-up points for the air removal system should be located above the vehicles near the ceiling in areas where CNG vehicles will be serviced.</u> Additional ventilation, controlled by a methane detection system, may be needed to safely handle large fuel releases.

• Based on NEC requirements for conventional fuels, all electrical systems both above and within 18 in. below the ventilation systems air inlet should meet NEC's Class 1 Division 2 rating.

- No heating unit, with open flames or surface temperatures at or above the ignition point of natural gas (~1000°F), should be installed in any area where the electrical recommendations require Division 2 wiring.

The code's recommended distance requirements (15 ft and 18 in.) are based on the diffusion properties of lighter than air flammable gases (similar to natural gas) and the properties of conventional fuel vapors, and may need to be modified based on vehicle and building specifics.

Fire Protection

NFPA 52 has a minimum requirement for a portable 20-B:C fire extinguisher to be present at the refueling area [5.4]. In addition, the vehicle engine must be turned off, signs must be present with the words "STOP MOTOR," "NO SMOKING," and "FLAMMABLE GAS," and no ignition sources shall be present within 3 m (10 ft) of the refueling connection to the vehicle. Shut-down switches for the compressor should be present both at the dispenser and at a location near the refueling site.

Other fire protection features have been incorporated into CNG refueling facilities such as methane detectors to warn of leaks from the dispenser, and automated fire suppression systems activated by ultraviolet/infrared detectors. Dry chemical is the preferred fire suppression material since water line protection from freezing is difficult in outdoor settings. The methane detectors can also be used to shut down the compressor and dispenser if desired.

Liquefied Natural Gas

Two NFPA standards govern LNG storage and dispensing into vehicles: NFPA 59A, *Production, Storage, and Handling of LNG* [5.16]; and NFPA 57, *Standard for Liquefied Natural Gas Vehicular Fuel Systems* [5.17]. These two standards are the most appropriate for designing and installing LNG vehicle storage and dispensing systems. NFPA 59A is targeted at large LNG facilities, 266,000 L (70,000 gal) or more, but includes basic methods of equipment fabrication and installation as well as operating practices for protection of persons and property. It addresses not only storage of LNG but production facilities as well. NFPA 57 is targeted to the design, installation, operation, and maintenance of LNG fuel systems on vehicles and their associated fuel storage and dispensing systems.

Neither NFPA 59A nor 57 deals with facilities that store or maintain LNG vehicles. For guidance on LNG vehicle storage and maintenance facilities, the NFPA standards 30 (*Flammable and Combustible Liquids Code*), 30A (*Automotive and Marine Service Station Code*), 52 (*Compressed Natural Gas Vehicular Fuel Systems*), 88A (*Standard for Parking Structures*), and 88B (*Repair Garages*) are of some assistance, though none specifically addresses LNG. The hazards posed by LNG vehicles indoors are similar to those posed by CNG vehicles with two very important exceptions: first, LNG is not odorized[1]; and, second, LNG may leak from the vehicle as a liquid and create vapor that is initially heavier than air until it warms sufficiently. It is important that these potential safety hazards be addressed from the perspective of safe facilities and to demonstrate to local fire and building officials that suitable precautions are being taken.

Location of Storage Tanks

According to NFPA 57, the location of LNG tanks is dependent on the size and location of the impoundment or containment system[2] for that tank. For tanks that are between 475 and 1900 L in size (125-500 gal), the minimum distance from property lines and buildings shall be at least 3 m (10 ft). For tanks of 114,000-266,000 L (30,001-70,000 gal), the distance shall be 23 m (75 ft). Tanks of sizes intermediate to these have minimum distances between these two limits.

NFPA 59A has the same distance guidelines as NFPA 57 for tanks up to 266,000 L (70,000 gal). For tanks greater than 266,000 L (70,000 gal), the minimum distance to property lines or buildings shall be 0.7 times the container diameter, but not less than 30 m (100 ft). NFPA 59A also requires that the impoundment area be located to prevent the heat flux from a fire from exceeding certain levels at the property line. The model used to calculate heat flux is described in the Gas Research Institute report GRI 0176, "LNGFIRE: A Thermal Radiation Model for LNG Fires" [5.18]. NFPA 59A also recommends that provisions be made to minimize the chance that flammable vapor clouds produced from an LNG spill will reach the property line. The flammable vapor cloud

---

[1] Attempts to find a suitable odorant for LNG are ongoing, but no practical odorant has been identified to date.

[2] Impoundment or containment systems are installed under and around tanks to contain any spills from that tank. Dikes are a typical example.

dispersion distances are to be calculated using the model described in the Gas Research Institute report GRI 0242, "LNG Vapor Dispersion Prediction with the DEGADIS Dense Gas Dispersion Model" [5.19]. While these models are intended for use where LNG is stored in large tanks, local fire officials may ask that these models be used regardless of the storage tank size.

Location and Installation of Fuel Dispensers

According to NFPA 57, LNG fuel dispensers shall be located no closer than 7.6 m (25 ft) from any property line or important building not associated with the LNG storage facility. The storage and dispensing facility must not be located where overhead electrical lines over 600 V might fall over them, and fired equipment must be kept away by the same distances as impoundment areas from property lines and buildings.

NFPA 57 recommends that LNG dispensers be protected from collisions with vehicles and that they incorporate an emergency shut-down system. The maximum delivery pressure of the dispenser shall not exceed the operating pressure of the LNG tanks on the vehicle being refueled, and the LNG delivery hoses must have a shut-off valve at the end and a breakaway valve in the event a vehicle drives off with the refueling hoses attached. Bleed and vent valve must be incorporated into the dispensing lines to allow them to be drained and depressurized prior to disconnection if necessary.

NFPA 57 addresses indoor refueling of LNG vehicles. The building housing the LNG dispenser shall have blowout panels in the roof or an exterior wall, and must have forced ventilation of at least five air changes per hour. The ventilation system need not operate continuously if activated by a methane detection system at one-fifth the lower flammability limit of methane. In either case, the dispensing system shall be interlocked with the ventilation system so that dispensing is allowed only when the ventilation system is operable. In addition, a methane detection system is required to shut down the refueling system when the methane content exceeds one-fifth of the lower flammability limit of methane. The methane detection system shall also provide audible and visual alarms when the methane content exceeds one-fifth of the lower flammability limit of methane. Reactivation of the refueling system following alarm by the methane detector shall be only through manual means by trained personnel.

Vehicle Storage and Maintenance Facilities

The safety hazard presented by LNG vehicles indoors is primarily one of fuel system leaks. Though LNG fuel systems operate at much lower pressure than CNG fuel systems, there is always the possibility that natural gas vapor will be released due to pressure build-up ("venting") or due to leaks. Pressure relief valves are required on LNG tanks to prevent excessive pressure from building up and rupturing the tank. Though LNG tanks have very good insulation, it is not perfect and all LNG tanks have a constant, though small, influx of heat into the LNG. This heat influx causes the pressure in the LNG tank to increase over time. When the pressure reaches the set point of the pressure relief valve, natural gas is vented until the pressure decreases to some preset value lower than the set relief pressure. In typical operation, an LNG vehicle will not be venting any natural gas because using LNG from tank causes the pressure to decrease, and because every time LNG is added to the tank it cools the LNG remaining in the tank. Venting should occur only if the vehicle is idle for an extended period of time or if the LNG tank has lost its vacuum or has some other problem degrading its insulating capability [5.20].

Most LNG vehicles provide a manifold to capture natural gas vented from the LNG tank and direct it to a safe location outside of the vehicle, usually near the top of the vehicle where the vented natural gas will rise and dissipate safely. For vehicle maintenance facilities, exhaust systems with explosion-proof blowers and motors should be provided. When the vehicle is brought inside for maintenance, the exhaust system hose is placed over the LNG vent and the system is turned on. Any vented natural gas vapor is then safely removed from the building. For vehicle parking facilities, this is usually not a practical solution for cost and implementation reasons.

Leaks from LNG vehicles cause similar safety hazards to those presented by leaks from CNG systems with the additional hazard that LNG vapor is not odorized and will not be detected without the use of methane detectors. Though it hasn't been proven with certainty, LNG fuel system fittings can be loosened by the thermal cycling they undergo. Compression fittings are preferred in place of flanged and threaded fittings for use in LNG fuel systems for this reason [5.21]. Other than these differences, the precautions that should be taken for buildings that store or maintain LNG vehicles are the same as for CNG vehicles. Please see the CNG Vehicle Storage and Maintenance Facilities section for details.

Fire Protection

Fire protection of LNG storage and dispensing systems should start with the minimum requirements in NFPA 52 for CNG compression and dispensing systems [5.4]:

- A portable 20-B:C fire extinguisher shall be present at the refueling area.

- The vehicle engine must be turned off when refueling is in operation.

- Signs must be present with the words "STOP MOTOR," "NO SMOKING," and "FLAMMABLE GAS."

- No ignition sources shall be present within 3 m (10 ft) of the refueling connection to the vehicle.

In addition, LNG storage and dispensing systems shall be protected by methane detectors that will sound an alarm and shut down the dispenser. A fire suppression system activated by infrared and ultraviolet flame sensors shall be included to protect the area where the vehicles are refueled. Additional measures such as automatically calling the fire department when the fire suppression system is activated should be considered.

LP Gas

The regulations governing LP gas fuel storage tank and fuel dispenser placement are complex and detailed. NPFA 58—*Standard for the Storage and Handling of Liquefied Petroleum Gases* [5.13] is the acknowledged "bible" of LP gas regulations and is accepted by most state and local building codes, though variations specific to certain areas are common. The following summarizes the requirements of NFPA 58, but it is not meant to be an exhaustive guide to meeting all NFPA 58 requirements. Local jurisdictions having authority may also adopt modifications to NPFA 58 based on local experience or preferences.

Location of Storage Tanks

All LP gas storage tanks shall be mounted outside of any buildings. Underground LP gas storage tanks shall not be installed under any buildings. The minimum distance that above-ground or underground LP gas storage tanks shall be installed from buildings or property lines depends on the size of the tanks. For underground tanks of 7600 L (2000 gal) or less, the minimum separation shall be 3 m or 10 ft. For larger underground tanks, the minimum distance shall be 15 m or 50 ft. For above-ground tanks of 1900 L (500 gal) or smaller, the minimum distance shall be 3 m or 10 ft. For above-ground tanks of 1900-7600 L (501-2000 gal) the minimum distance shall be 7.6 m or 25 ft. For above-ground tanks of 7600-114,000 L (2001-30,000 gal), the minimum distance shall be 15 m or 50 ft. Above-ground LP gas tanks must be placed at least 6 m (20 ft) from any other above-ground tank containing liquids having flash points less than 93.4°C (200°F).

Location and Installation of Fuel Dispensers

LP gas fuel dispensers shall be mounted on a concrete base or attached directly to a refueling unit. Dispensers must be mounted away from buildings, property lines, public ways, driveways, above-ground tanks of other flammable liquids, sewer system openings, etc., by the distances listed in NFPA 58. Dispensing hoses may not be longer than 5.5 m (18 ft) unless approved by local jurisdictions. The dispensing hose must contain an emergency breakaway device that contains liquid on both sides to guard against vehicles driving off with the dispensing hose still connected. The dispenser must contain an excess flow valve or an emergency shut-off valve where the dispensing hose connects to the rigid piping. LP gas fuel dispensers may be placed under canopies or other means of providing shelter from the elements, but the perimeter of the dispenser must not be enclosed by walls by more than 50%.

Vehicle Storage and Maintenance Facilities

Buildings designed for gasoline and diesel vehicles will accommodate LP gas vehicles safely as well. NFPA 58 does provide some additional cautions when LP gas vehicles are brought indoors. First, the fuel system must be leak-free and the storage tank(s) shall not be filled beyond its capacity. Second, when vehicles are under repair, the container shut-off valve shall be closed except when it is necessary to operate the engine. Third, LP gas vehicles shall not be parked near sources of heat, open flames, or similar sources of ignition, or near inadequately ventilated pits.

Fire Protection

A fire suppression system should be provided around the dispenser to protect personnel and vehicles from fire during refueling. Actuation shall be provided both by sensors automatically and through the use of a manual switch. LP gas tanks that can be exposed to fire should have water spray fixed systems to prevent the tanks from failure due to over-temperature. The direct application of water in the form of a spray can also be used to control unignited gas leakage.

Vegetable Oils

To date, the vegetable oils that have been seriously considered for use in highway transportation vehicles have been esters of vegetable oils mixed with diesel fuel in percentages of 20% or 30%. However, esters of vegetable oils themselves have very similar properties to diesel fuel, and there is no reason that they would be treated differently from diesel fuel in terms of building codes.

Location of Storage Tanks

Storage tanks containing mixtures of vegetable oils and diesel fuel, vegetable oils, or esters of vegetable oils should be allowed to be placed wherever diesel fuel tanks are placed. While vegetable oils or esters of vegetable oils are biodegradable, the same tank leak detection and prevention practices should be followed as for diesel fuel to avoid loss of product and contamination of the ground water.

Location and Installation of Fuel Dispensers

Fuel dispensers for vegetable oils and esters of vegetable oils should be allowed to be placed wherever dispensers for diesel fuel are allowed to be placed.

Vehicle Storage and Maintenance Facilities

No changes in vehicle storage and maintenance facilities should be required for vegetable oils or esters of vegetable oils. Vegetable oil esters may have small amounts of methanol or ethanol (depending on the type of ester) present in them which could potentially raise a problem with oil and water separators if large amounts of fuel are ever spilled in these facilities and the water discharged from these facilities is not treated. Spills from refueling or normal maintenance should not cause any concern since it is likely that the methanol or ethanol would vaporize before entering the oil and water separator.

Fire Protection

The same fire protection required for diesel fuel should be used for vegetable oils and esters of vegetable oils. Dry chemical fire suppression systems are the most appropriate for use at vehicle fuel storage and dispensing systems. The discharge nozzles should be placed both overhead and at curb level so that the entire area prescribed by the radius of the dispensing hose will be covered. Testing of the nozzle placement should be done by using a "puff test" in which the dry chemical is discharged on purpose to verify coverage. (The puff test need not discharge the entire amount of dry chemical, just enough so that the area of coverage can be determined with certainty.) The puff test is triggered by creating a small controlled fire of the fuel used which also tests the fire detection sensors, usually ultraviolet and infrared.

Hydrogen

Building codes specifically addressing hydrogen storage and dispensing into vehicles do not exist. If use of hydrogen as an alternative fuel becomes more likely in the future, it is probable that codes and standards specific to hydrogen storage and dispensing for vehicle use will be developed. At present, NFPA 52 represents the standard most similar to what might be developed for compressed hydrogen storage and dispensing systems. NFPA 57 represents the standard most similar to what might be developed for liquefied hydrogen storage and dispensing systems. In the interim, these standards, along with a thorough understanding of safe hydrogen handling practices, might allow building of safe vehicular storage and dispensing systems.

Electricity

The electric vehicle (EV) industry is working to develop comprehensive building codes through the National Electrical Code (NEC). What has been developed to date are codes for sizing the electrical service for EV chargers, EV connectors between the vehicle and the charger, and preliminary requirements for ventilation in charging areas to prevent a potentially hazardous build-up of hydrogen released from the batteries during recharging. All of these requirements are included in NEC Article 625 that was published in the 1996 NEC [5.22]. The advanced batteries being developed for EVs may not release any hydrogen at all which would alleviate the concern about ventilation. Figure 5-3 illustrates an EV charger used to recharge electric transit buses.

Sizing of Charger Service

The electrical service for EV chargers must be sized to accommodate 125% of the rated maximum power of the charger, assuming that load is continuous. Elec-

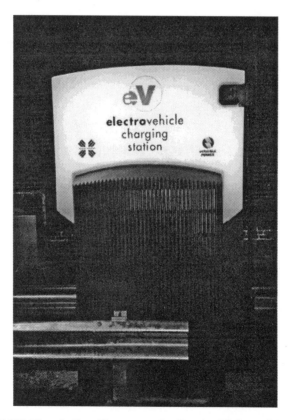

Fig. 5-3 25 kW Electric Bus Charger at the Greater Richmond Transit Company.

trical wire sizes and conduit selection must be done in accordance with the applicable NEC codes.

Location of Chargers

The EV chargers currently being developed are designed for use both inside and outside buildings. When mounted outside, some sort of overhead protection such as a canopy is recommended to provide protection from rain and snow as is the current practice for gasoline and diesel fuel refueling dispensers. EV chargers do generate heat when in operation and most have some sort of cooling system. Adequate ventilation must be provided to prevent overheating of the chargers, though this is a concern only if the chargers are placed in small, enclosed spaces. Bollards and guardrails should be provided to prevent vehicles from inadvertently running into the chargers when maneuvering in position to be recharged. Chargers located in flood zones must be located above the base flood elevation or waterproofed, and have ground fault protection as required by existing codes for any electrical equipment that may become submerged.

NEC 625 requires that the chargers must be located so that the charging cable can be directly connected to the EV. The recharging connector of the EV must be located within a zone 0.46-1.22 m (18-48 in.) above the floor. Both conductive and inductive recharging connectors are allowed. Figure 5-4 illustrates an electric bus being recharged using an inductive paddle recharging connector.

Vehicle Charging Facilities

The safety concerns for vehicle charging facilities include hydrogen produced by the batteries during vehicle recharging and spills of battery electrolytes. Ignition hazards presented by the charging equipment are also a concern if other hazardous materials are present such as conventional vehicle fuels, solvents, etc.

EV battery technology is evolving quickly, with more advanced designs becoming more numerous. Some of these advanced designs do not emit hydrogen during charging, but several do. Lead-acid batteries (flooded and sealed) typically do emit hydrogen during charging, but the amount varies significantly with battery model and manufacturer. For recharging EVs inside, forced ventilation is required

Fig. 5-4 Electric Bus Being Recharged Using an Inductive Recharging Connector.

if the batteries are capable of releasing hydrogen. NEC 625 states the requirements under which forced ventilation must be used and provides guidance for estimating the level of ventilation required. Forced ventilation, if required, must be permanently installed, interlocked with operation of the chargers, and sufficient to keep the hydrogen concentration below 1% during vehicle charging (which is 25% of the lower flammability limit of 4% hydrogen in air).

If the EV has been certified to not release hydrogen during charging, forced ventilation is not required. The Society of Automotive Engineers (SAE) has developed a Recommended Practice, J1718, "Measurement of Hydrogen Gas Emission from Battery-Powered Passenger Cars and Light Trucks During Battery Charging," to determine whether hydrogen emission from an EV is small enough to not require forced ventilation [5.23]. The tests of J1718 include simulation of normal charging and potential equipment failure modes. Chargers, receptacles, and power outlets meant for charging EVs that do not require forced ventilation must be clearly marked to prevent charging of EVs that do require forced ventilation.

Some battery designs have liquid electrolyte in them that would have to be treated as hazardous materials if they were released. However, the actual quantities of such materials in batteries is very small and the probability of release during charging is remote. The most likely concern about release of battery electrolyte would be during vehicle collisions where the batteries are physically damaged.

Fire Protection

Facilities where EVs are charged or stored should be protected from fire. The same fire protection technologies used for conventional vehicles should be used for EVs. In charging facilities, sprinkler systems can be used because chargers are designed to withstand the water exposure that a sprinkler system would cause. Smoke alarms and ultraviolet/infrared detectors are recommended for EVs, just as they would be for conventional vehicles.

Sources of Additional Information

The following reports and organizations have additional and more detailed information about alternative fuel safety, refueling facility installation, and facility modifications to safely store and maintain alternative fuel vehicles.

- For all alternative fuels:

Garage Guidelines for Alternative Fuels, March 1995
The New York State Energy Research and Development Authority
Corporate Plaza West
286 Washington Avenue Extension
Albany, New York 12203-6399
Phone: 518-862-1090
Fax: 518-862-1091
http://www.nyserda.org

Resource Guide—Infrastructure for Alternative Fuel Vehicles, June 1995
The California Energy Commission
Transportation Technology & Fuels
1516 Ninth Street, MS-41
Sacramento, California 95814
Phone: 916-654-4634
http://www.energy.ca.gov

Safe Operating Procedures for Alternative Fuel Buses, 1993
Transportation Research Board
National Research Council
2101 Constitution Avenue, N.W.
Washington, D.C. 20418

Summary Assessment of the Safety, Health, Environmental and System Risks of Alternative Fuels, August 1995
U.S. Department of Transportation
Research and Special Programs Administration
John A. Volpe National Transportation Systems Center
Cambridge, Massachusetts 02142-1093

"Addressing the Fire Hazards of Alternative Fuels for Public Transit Buses"
Industrial Fire Safety (Part 1, September/October 1993; Part 2, January/February 1994)

National Fire Protection Handbook
National Fire Protection Association
1 Batterymarch Park
P.O. Box 9101
Quincy, Massachusetts 02269-9101
Phone: 800-735-0100

- For methanol:

Machiele, P.A., "Summary of the Fire Safety Impacts of Methanol as a Trans-portation Fuel," SAE Paper No. 901113
Society of Automotive Engineers
400 Commonwealth Drive
Warrendale, Pennsylvania 15096-0001
Phone: 412-776-4841

Petroleum Marketing Fire-Safety Considerations for Alcohols, Ethers, and Gasoline Alcohol and Ether Blends, May 1995
American Petroleum Institute
1220 L Street, N.W.
Washington, D.C. 20005

Alcohols and Ethers—A Technical Assessment of Their Application as Fuels and Fuel Components, July 1988
American Petroleum Institute
1220 L Street, N.W.
Washington, D.C. 20005

Storage and Handling of Gasoline-Methanol/Cosolvent Blends at Distribution Terminals and Service Stations, API Recommended Practice 1627, August 1986
American Petroleum Institute
1220 L Street, N.W.
Washington, D.C. 20005

Design Guidelines for Bus Transit Systems Using Alcohol Fuel (Methanol and Ethanol) as an Alternative Fuel, August 1996
U.S. Department of Transportation
Research and Special Programs Administration
John A. Volpe National Transportation Systems Center
Cambridge, Massachusetts 02142-1093

- For ethanol:

Fuel Ethanol—Technical Bulletin, September 1993
Archer Daniels Midland Company
P.O. Box 1470
Decatur, Illinois 62525

Storing and Handling Ethanol and Gasoline-Ethanol Blends at Distribution Terminals and Service Stations, API Recommended Practice 1626, April 1985
American Petroleum Institute
1220 L Street, N.W.
Washington, D.C. 20005

Petroleum Marketing Fire-Safety Considerations for Alcohols, Ethers, and Gasoline Alcohol and Ether Blends, May 1995
American Petroleum Institute
1220 L Street, N.W.
Washington, D.C. 20005

Design Guidelines for Bus Transit Systems Using Alcohol Fuel (Methanol and Ethanol) as an Alternative Fuel, August 1996
U.S. Department of Transportation
Research and Special Programs Administration
John A. Volpe National Transportation Systems Center
Cambridge, Massachusetts 02142-1093

National Corn Growers Association
1000 Executive Parkway, Suite 105
St. Louis, Missouri 63141
Phone: 314-275-9915

National Ethanol Vehicle Coalition
1648 Highway 179
Jefferson City, Missouri 65109
Phone: 314-635-8445

Guidebook for Handling, Storing, & Dispensing Fuel Ethanol
U.S. Department of Energy
Alternative Fuels Data Center
http://www.afdc.doe.gov

- For compressed and liquefied natural gas:

Design Guidelines for Bus Transit Systems Using Compressed Natural Gas as an Alternative Fuel, June 1996
U.S. Department of Transportation
Research and Special Programs Administration
John A. Volpe National Transportation Systems Center
Cambridge, Massachusetts 02142-1093

Liquefied Natural Gas Safety in Transit Operations, March 1996
U.S. Department of Transportation
Research and Special Programs Administration
John A. Volpe National Transportation Systems Center
Cambridge, Massachusetts 02142-1093

NFPA 52—Standard for Compressed Natural Gas (CNG) Vehicular Fuel Systems 1995 Edition
National Fire Protection Association
1 Batterymarch Park
P.O. Box 9101
Quincy, Massachusetts 02269-9101
Phone: 800-735-0100

Compressed Natural Gas Safety in Transit Operations, October 1995
U.S. Department of Transportation
Research and Special Programs Administration
John A. Volpe National Transportation Systems Center
Cambridge, Massachusetts 02142-1093

Codes and Standards Applicable to Natural Gas Vehicle Fuel Stations Primarily for Compressed Natural Gas, February 1990
Natural Gas Vehicle Coalition
1515 Wilson Boulevard, Suite 1030
Arlington, Virginia 22209
Phone: 703-527-3022
Fax: 703-527-3025

Compressed Natural Gas Vehicle (NGV) Fueling Connection Devices, 1994
American National Standard/Canadian Gas Association Standard
American Gas Association
8501 East Pleasant Valley Road
Cleveland, Ohio 44131

Liquefied Natural Gas Vehicle Applications
LNG Vehicle Markets and Infrastructure, September 1995
An Introduction to LNG Vehicle Safety, December 1991
LNGFIRE: A Thermal Radiation Model for LNG Fires
LNG Vapor Dispersion Prediction with the DEGADIS Dense Gas Dispersion Model
Gas Research Institute
8600 West Bryn Mawr Avenue
Chicago, Illinois 60631-3562
Phone: 312-399-8100
Fax: 312-399-8170
http://www.gri.org

* For propane:

NFPA 58—Standard for the Storage and Handling of Liquefied Petroleum Gases, 1995 Edition
National Fire Protection Association
1 Batterymarch Park
P.O. Box 9101
Quincy, Massachusetts 02269-9101

U.S. Department of Transportation
Research and Special Programs Administration
John A. Volpe National Transportation Systems Center
Cambridge, Massachusetts 02142-1093

National Propane Gas Association
1600 Eisenhower Lane
Lisle, Illinois 60532
Phone: 708-515-0600
Fax: 708-515-8774

Propane Vehicle Council
1101 17th Street, N.W., Suite 1004
Washington, D.C. 20036
Phone: 202-530-0479
Fax: 202-466-7205

- For electricity:

    National Electrical Code Handbook, Seventh Edition, March 25, 1996
    National Fire Protection Association
    1 Batterymarch Park
    P.O. Box 9101
    Quincy, Massachusetts 02269-9101

    Electric Transportation Coalition
    701 Pennsylvania Avenue, N.W., 4th Floor
    Washington, D.C. 20004
    Phone: 202-508-5995

    Electric Vehicle Association of the Americas
    601 California Street, Suite 502
    San Francisco, California 94108
    Phone: 408-253-5262

Edison Electric Institute
701 Pennsylvania Avenue, N.W., 4th Floor
Washington, D.C. 20004
Phone: 202-508-5000

Electric Power Research Institute
P.O. Box 10412
Palo Alto, California 94303
Phone: 415-855-2984

Electric Vehicle Industry Association
P.O. Box 85905
Tucson, Arizona 85754
Phone: 602-889-0248

• For biodiesel:

National BioDiesel Board
2405 Grand, Suite 700
Kansas City, Missouri 64108
Phone: 816-474-9407

National SoyDiesel Development Board
P.O. Box 194898
Jefferson City, Missouri 65110-4898
Phone: 800-841-5849

References

5.1. Geyer, W.B., P.E., P.O.E., "A Roadmap of AST Codes and Standards," *Petroleum Equipment and Technology*, January/February 1997, Vol. 2, No. 1, HMS Publishing, Inc., Bamington, Ill.

5.2. *Flammable and Combustible Liquids Code—1993 Edition (NFPA 30)*, National Fire Protection Association, Quincy, Mass.

5.3. *Automobile and Marine Service Station Code—1993 Edition (NFPA 30A)*, National Fire Protection Association, Quincy, Mass.

5.4. *Standard for Compressed Natural Gas (CNG) Vehicular Fuel Systems— 1995 Edition (NFPA 52)*, National Fire Protection Association, Quincy, Mass.

5.5. *National Electrical Code—1993 Edition (NFPA 70)*, National Fire Protection Association, Quincy, Mass.

5.6. *Design Guidelines for Bus Transit Systems Using Compressed Natural Gas as an Alternative Fuel*, U.S. Department of Transportation, Federal Transit Administration, DOT-FTA-MA-26-7021-96-1, June 1996.

5.7. Moog, C.G., and Bechtold, R.L., "Safety Issues Raised by Repair of CNG Vehicles at the SUNY Buffalo North Campus Vehicle Maintenance Facility," January 15, 1995, New York State Energy Research and Development Authority.

5.8. National Fire Protection Association, *NFPA 88B Standard for Repair Garages—1985 Edition*, Section 3-3.3.

5.9. Ibid. Section 3-2.3.1.

5.10. Ibid. Section 3-2.2.2.

5.11. National Electrical Code, Articie 511-3a.

5.12. *Recommended Practice for Classification of Class I Hazardous (Classified) Locations for Electrical Installations in Chemical Process Areas (NFPA 497A)*, National Fire Protection Association, Quincy, Mass.

5.13. *Standard for the Storage and Handling of Liquefied Petroleum Gases— 1995 Edition (NFPA 58)*, National Fire Protection Association, Quincy, Mass., February 7, 1995.

5.14. Murphy, M., *et al.*, "Extent of Indoor Flammable Plumes Resulting from CNG Bus Fuel Leaks," SAE Paper No. 922486, Society of Automotive Engineers, Warrendale, Pa., 1992.

5.15. Grant, T., *et al.*, "Hazard Assessment of Natural Gas Vehicles in Public Parking Garages," Ebasco Services Incorporated, July 1991.

5.16. *Production, Storage, and Handling of Liquefied Natural Gas (LNG)—1996 Edition (NFPA 59A)*, National Fire Protection Association, Quincy, Mass.

5.17. *Standard for Liquefied Natural Gas Vehicular Fuel Systems—1996 Edition (NFPA 57)*, National Fire Protection Association, Quincy, Mass.

5.18. "LNGFIRE: A Thermal Radiation Model for LNG Fires," Report GRI 0176, Gas Research Institute, Chicago, Ill.

5.19. "LNG Vapor Dispersion Prediction with the DEGADIS Dense Gas Dispersion Model," Report GRI 0242, Gas Research Institute, Chicago, Ill.

5.20. Gibbs, J.L., Bechtold, R.L., and Collison, C.E. III, "The Effects of LNG Weathering on Fuel Composition and Vehicle Management Techniques," SAE Paper No. 952607, Society of Automotive Engineers, Warrendale, Pa., 1995.

5.21. Midgett, Dan E. II, "Best Available Practices for LNG Fueling of Fleet Vehicles," Topical Report No. GRI-96/0180, Gas Research Institute, Chicago, Ill., February 1996.

5.22. National Electrical Code NEC 625.

5.23. "Measurement of Hydrogen Gas Emission from Battery-Powered Passenger Cars and Light Trucks During Battery Charging," SAE J1718, Society of Automotive Engineers, Warrendale, Pa., December 1994.

Chapter Six

Glossary of Terms

acetaldehyde	Toxic emission produced from combustion; a combustion intermediate of ethanol.
ASME	American Society of Mechanical Engineers—maintains a pressure vessel code that propane and CNG cylinders can be built to meet.
ASTM	American Society for Testing and Materials—certifies standard tests of fuel properties.
autoignition temperature	Minimum temperature of a substance to initiate self-sustained combustion independent of any ignition source.
bifuel	Vehicles with two fuel systems, but with only one useable at a time.
biodiesel	Generic descriptor for esterified vegetable oils, also used to describe blends of esterified vegetable oils with diesel fuel, e.g., B20 denotes 20 volume percent esterfied vegetable oil in diesel fuel.
BOCA	Building Officials and Code Administration.

boiling temperature | Temperature at which the transformation from liquid to vapor phase occurs in a substance at a pressure of 14.7 psi (atmospheric pressure at sea level). Fuels that are pure compounds (such as methanol) have a single temperature as their boiling points, while fuels with mixtures of several compounds (like gasoline) have a boiling temperature range representing the boiling points of each individual compound in the mixture. For these mixtures, the 10% point of distillation is often used as the boiling point.

building code | Regulation for building design and construction enforced by a local jurisdiction, usually the fire marshall.

California Phase 2 gasoline | Specially reformulated gasoline that must be used in California to reduce emissions.

cascade | Compressed natural gas storage system consisting of three or more cylinders allowing fast-fill of compressed natural gas vehicles.

cetane number | The ignition quality of a diesel fuel measured using an engine test specified in ASTM Method D 613. Cetane number is determined using two pure hydrocarbon reference fuels: cetane which has a cetane rating of 100; and heptamethylnonane (also called isocetane), which has a cetane rating of 15.

CNG | Compressed natural gas—natural gas compressed for use in vehicles.

CO | Carbon monoxide—exhaust emission resulting from incomplete combustion and/or fuel-rich combustion.

conservation vent	A valve on fuel storage systems that controls both the pressure and vacuum allowed to build up inside fuel storage tanks. Used with low vapor pressure fuels such as gasoline, methanol, and ethanol.
cryogenic	Refers to very low temperatures (typically liquids with boiling points $\leq -100°C$ [$-148°F$]).
density	Mass per unit volume, expressed in kg/L or lb/gal.
DME	Dimethyl ether—an alternative fuel for diesel engines made from natural gas.
dry-break	Refueling connection that closes automatically when disconnected.
E85	A blend of 85 volume percent ethanol and 15 volume percent gasoline—the gasoline provides cold-start and safety benefits.
elastomer	Used to describe a wide range of "soft" materials used in fuel systems.
electrical conductivity	Measure of the ability of a substance to conduct an electrical charge.
EPA	U.S. Environmental Protection Agency—sets vehicle emissions regulations and fuel storage leakage requirements.
ETBE	Ethyl tertiary butyl ether—an ether made from ethanol and iso-butylene that can be used in reformulated and oxygenated gasolines.

ethanol	Alcohol made from agricultural crops or bio-mass used as an alternative fuel in spark-ignition and diesel engines.
EV	Electric vehicle.
fast-fill	CNG refueling systems that allow refueling in about the same time as a typical gasoline vehicle.
FFV	Flexible fuel vehicle—vehicles able to operate on gasoline, M85 or E85, or any mixture of the two gasoline and alcohol fuels.
flame spread rate	Rate of flame propagation across a fuel pool.
flame visibility	Degree to which combustion of a substance under various conditions can be seen.
flammability limits	Minimum and maximum concentrations of vapor in air below and above which the mixtures are unignitable. A vapor-air concentration below the lower flammable limit is too lean to ignite, while a concentration above the upper flammable limit is too rich to ignite.
flammable	Capable of supporting combustion.
flashback	A backfire through the intake system of an engine using hydrogen.
flash point	Minimum temperature of a liquid at which sufficient vapor is produced to form a flammable mixture with air.
formaldehyde	Toxic emission produced from combustion; a combustion intermediate of methanol.

freezing point	The temperature where a liquid can exist as both a liquid and a solid in equilibrium.
gasohol	A 10 volume percent blend of ethanol with gasoline.
global warming	Theory that the earth's atmosphere is gradually getting warmer, primarily due to fossil fuel combustion.
HC	Hydrocarbon emissions.
HD-5	Propane specification that requires 90% propane minimum, 2.5% butane maximum, and 5% propylene maximum.
heating value	The heat released when a fuel is combusted completely, corrected to standard pressure and temperature. The higher heating value is complete combustion with the water vapor in the exhaust gases condensed. The lower heating value is when the water vapor in the exhaust is in the vapor phase.
hold time	Time from when an LNG tank is filled to when pressure build-up requires vapor to be vented.
IFCI	International Fire Code Institute.
Joule-Thompson cooling	The decrease in temperature of a gas when its pressure is rapidly reduced.
latent heat of vaporization	The quantity of heat absorbed by a fuel in passing between liquid and gaseous phases. The conditions under which latent heat of vaporization is measured is the boiling point (or boiling range) and atmospheric pressure, 101.4 kPa (14.7 psi).

LEV	Low Emission Vehicle—the second most-stringent California vehicle emission standard for combustion vehicles.
LFL	Lower flammability limit—the least amount of fuel in air that will support combustion.
LH_2	Liquefied hydrogen.
LNG	Liquefied natural gas—natural gas liquefied for use in vehicles.
LP gas	Liquefied petroleum gas; most typical example is propane, but butane and ethane are other common LP gases.
M100	Used to represent pure, 100% or "neat" methanol.
M85	A blend of 85 volume percent methanol and 15 volume percent gasoline—the gasoline provides cold-start and safety benefits.
methane	The lightest hydrocarbon, and primary constituent of natural gas.
methanol	Alcohol made from natural gas used as an alternative fuel in spark-ignition and diesel engines.
molecular weight	The sum of the atomic weights of all the atoms in a molecule.
MTBE	Methyl tertiary butyl ether—an ether made from methanol and iso-butylene that is used widely in reformulated and oxygenated gasolines.
NEC	National Electrical Code.

NFPA	National Fire Protection Association—certifies building codes and standards.
NMOG	Non-methane organic gases—all organic gases in vehicle exhaust not counting methane.
NO_x	Oxides of nitrogen—exhaust emission created from reactions of nitrogen in the air with oxygen at peak combustion temperatures.
octane number	A measure of the resistance of a fuel to combustion knock using standardized engine tests. The Research octane number is determined using ASTM Method D 2699; the Motor octane number is determined using ASTM Method D 2700. The Antiknock Index is the average of the Research and Motor numbers. Octane numbers are determined using n-heptane that has an octane number of 0, and isooctane that has an octane number of 100.
odor recognition	Degree of smell associated with fuel vapor.
oil-less compressors	Compressors specifically designed to minimize oil in the cylinders; designed specifically for compressing natural gas to prevent oil contamination.
OMHCE	Organic matter hydrocarbon equivalent—a weighted measure of hydrocarbon emissions based on the reactivity of all the components.
plasticizer	Generic term to describe components in elastomers added to increase suppleness and deformability.
PRD	Pressure relief device—used on pressure vessels to prevent failure due to over-pressure.

propane

The most common LP gas, and the most suitable for use as an alternative vehicle fuel.

puff test

A test of a dry chemical fire suppression system that determines whether adequate coverage of intended areas is achieved.

RVP

Reid vapor pressure (see "vapor pressure").

SAE

Society of Automotive Engineers.

SBCCI

Southern Building Code Congress International.

secondary containment

EPA requirement to catch leaks from underground storage tanks.

slow-fill

CNG refueling systems designed to refuel CNG vehicles over a long period of time (e.g., overnight).

SO_2

Sulfur dioxide—emission from oxidation of sulfur in the fuel.

specific gravity

The ratio of the density of a material to the density of water.

specific heat

The ratio of the heat needed to raise the temperature of a substance one degree compared to the heat needed to raise the same mass of water one degree.

specific reactivity

The ozone-forming potential per gram of exhaust or evaporative emissions.

stoichiometric air-fuel ratio

The exact air-fuel ratio required to completely combust a fuel.

TBA

Tertiary butyl alcohol—an oxygenate used for blending with gasoline.

TLEV

Transitional Low Emission Vehicle—the least-stringent California vehicle emission standard for combustion vehicles.

ULEV

Ultra Low Emission Vehicle—the most-stringent California vehicle emission standard for combustion vehicles.

UST

Underground storage tank.

vapor density

Weight of a volume of pure (no air present) vapor compared to the weight of an equal volume of dry air at the same temperature and pressure. A vapor density of less than one describes a vapor that is lighter than air, while a value greater than one describes a vapor that is heavier than air.

vapor pressure

Equilibrium pressure exerted by vapors over a liquid at a given temperature. The Reid vapor pressure (RVP) is typically used to describe the vapor pressure of petroleum fuels without oxygenates at 100°F (ASTM Test Method D 323, Test Method for Vapor Pressure of Petroleum Products). The term "true vapor pressure" is often used to distinguish between vapor pressure and Reid vapor pressure. The Reid vapor pressure test involves saturating the fuel with water before testing and cannot be used for gasoline-alcohol blends or neat alcohol fuels; a new procedure has been developed which does not use water and is called Dry Vapor Pressure Equivalent, or DVPE (see ASTM D 4814-95c, Standard Specification for Automotive Spark-Ignition Engine Fuel).

venting   Releasing vapor from a fuel tank to prevent over-
                                         pressurization.

viscosity   The resistance of a liquid to flow.

Viton   A trade name for a class of elastomers that will
                                         have differing responses to methanol and other
                                         fuels.

water solubility   Maximum concentration of a substance that will
                                         dissolve in water.

weathering   The loss of methane vapor and the deterioration
                                         of methane content in LNG.

Index

About the Author

Richard L. Bechtold is a registered professional engineer in Maryland with 20 years of experience with alternative fuels. His interest in this growing field was sparked by the gasoline shortage of 1973 to 1974.

Mr. Bechtold earned B.S. and M.S. degrees in mechanical engineering from Penn State University and prepared his master's thesis on the blending of used lubricating oil with diesel fuel to both extend the diesel fuel supply and remove a serious source of ground water pollution. Mr. Bechtold's interest in fuels other than gasoline and diesel fuel was augmented by the engine testing and exhaust gas analysis skills that he gained while obtaining his M.S. degree.

Mr. Bechtold's first job was as a mechanical engineer at the Bartlesville Energy Technology Center (BETC) in Bartlesville, Okla., from 1976 until 1980. There opportunities existed to conduct tests of alternative fuels on both engine and chassis dynamometers, modify vehicles to operate on alternative fuels, and conduct field tests of alternative fuel vehicles. Approximately a dozen professionals at BETC were engaged in alternative fuels research at that time, and the mix of skills, experience, and talents present at BETC provided a rare and excellent learning environment. It is interesting to note that this small group of professionals has excelled in industry and government positions since then.

Following his tenure at BETC, Mr. Bechtold accepted a position with Mueller and Associates, Inc., a small consulting engineering firm that specialized in providing technical expertise and skills about alternative transportation fuels. During this time, interest in alternative fuels grew dramatically throughout the United States. As the major auto manufacturers became more involved, the focus of interest began to shift from engine and vehicle research to field demonstrations of alternative fuel vehicles. Mr. Bechtold participated in and managed several of these demonstrations and gained notoriety in the alternative fuels community through the publication of numerous technical papers.

Mueller Associates was acquired by EA Engineering in 1988, where Mr. Bechtold now has the title of Sr. Project Manager.

More recently, the focus of interest has shifted to the infrastructure (i.e., fuel availability, storage, dispensing, and mechanic training) to support alternative fuel vehicles and innovative public policies to promote and sustain alternative fuel vehicles. The recently enacted Energy Policy Act (EPACT) of 1992 mandated Federal, state, and fuel provider fleets to purchase alternative fuel vehicles starting in 1996 and is the most ambitious U.S. legislation to promote alternative fuel vehicles to date. The decision to require private fleets to purchase alternative fuel vehicles is scheduled to be made by 2002. Many of the fleet managers acquiring alternative fuel vehicles to satisfy EPACT will also need to acquire alternative fuel storage and dispensing systems, and possibly modify their vehicle storage and maintenance facilities.

The impetus for this book was the knowledge that thousands of fleet managers and engineers would seek information about alternative fuels. Through this book, Mr. Bechtold shares his unique and valuable experience with readers in an effort to foster the development and understanding of alternative fuels use.